A CAREER
IN THE
PETROLEUM
INDUSTRY

In a Time of Energy Transition

Livinus Nosike, *PhD, mNAPE, mAAPG*

A CAREER IN THE PETROLEUM INDUSTRY

In a Time of Energy Transition

Delizon books and other media products are available at special quantity discounts, for use as premiums and sales promotion, or in corporate training programmes. For more information, please send your email to the Director of Sales: sales@delizonpublishers.com

Cover design by:

Delizon International Book Publishers

www.delizonpublishers.com

Europe   .   Africa   .   North America   .   Asia

Dedication

iii

*To Ify, my lovely wife,
for her patience and support during these years
on the path to success with many challenges.*

Acknowledgements

I acknowledge Dr Kingsley Ojoh, the late Executive Director of the Total Upstream Company in Nigeria, without whom I would never have gained the knowledge necessary to write this book. Not until after my own experiences of studying, training and working abroad for years, and coming back to the enclave of a workplace in our oil and gas sector, somewhat distanced from the masses, did I understand what it took for him to go so far, driving across many kilometres searching to recruit ordinary Nigerians without bias. He was a great mentor who worked hard to secure a merit-based international scholarship programme tailored to young Nigerian recruitment candidates. Also thanks to the Total Group for their sponsorship of my postgraduate studies and training.

I thank the many wonderful people I met in France and the many Nigerians living there at that time, who all contributed to the administrative processes and by way of moral support. It touches me when I see those passionate colleagues lamenting and wishing that things were better on this side, the African side of the globe. Some of them honestly looked forward to a better future for all. Thanks to the French government, which cosponsored my postgraduate scholarship with Total E&P (Operator of the Joint Venture with NNPC), and eased the administrative aspect of my studies through CNOUS, the French administrative body for scholars. I remember the salient words of my friends at *Ecole Doctorale*, University of Nice – Sophia Antipolis, especially international students, during our professional conferences and workshops. They looked up to me, asking if I could make a change back home. I saw how much the hearts of many aspired for a better world, where each recruit in the petroleum industry worked with conscience. I thank you all for your relevant contributions to this cause.

I must not forget to mention the staff and well wishers at Delizon Publishers, who in collaboration with Integrated Elvee Services Limited, worked to spread this goodwill message to the masses. Thanks to comrade Emmanuel Ikwueze and Mrs Jo Egrè, my longstanding editors. Thanks also to Prof. A. W. Mode, Prof. C. S. Nwajide, Dr Charles Ugbor, Dr Aaron Auduson and others who partner with me in fostering the career paths of many young candidates in the energy sector. With the collective efforts, we hope that this very society where we live and breathe, where we have been harboured and nurtured to greatness despite the numerous challenges, will see the light of a new perspective in the oil and gas industry.

Over the last few months, many of you who have read my book: *Overview of the Petroleum Industry,* have supported me by direct calls, contacts through our website, and mainly through the social media. You have been an enormous support and encouragement.

Thanks to all.

Disclaimer

This work provides a guide to a career in the petroleum industry, with its ups and downs. Any reference to individuals, persons or companies is aimed at giving a general opinion of oil and gas activity and not intended to offend any character.

Contents

Dedication...*iii*
Acknowledgements..*iv*
Disclaimer..*vi*
Contents..*vii*
Introduction...*viii*

Chapter One
CAREER: TYPES AND ACCESSIBILITY......................1

Chapter Two
EMPLOYMENT IN AN OIL AND GAS COMPANY......13

Chapter Three
JOB SECURITY AND WORK PRESSURE.................31

Chapter Four
HUMAN RESOURCES AND INTERNAL FUNCTIONS...59

Chapter Five
SELECTIVE RECRUITEMENT AND CAREER PROFILING..75

Chapter Six
HOW THE SYSTEM WORKS TO TAME YOU..........101

Chapter Seven
DEVELOPING OIL PRODUCING NATIONS............125

Chapter Eight
THE ENERGY INDUSTRY – A CHANGING CONCERN...157

Chapter Nine
EXIT STRATEGY...185

Afterword..*213*

Introduction

Some of us are familiar with the oil company towers and smart offices and the people working in those premises; then the technical settings, the exploration and production departments in these complexes, and the top management executives who appear at public conferences and workshops. Many are unaware of what really goes on in the daily work and life patterns of the employees pursuing a career in the oil and gas industry. The petroleum workspace has been a closed circle for many years now, with much intrigue and quiet absorption of its personnel. Even within the industry, not all the members of staff are involved directly in oil and gas activities, the rig sites, the onshore and offshore well operations and the flow stations. They get to understand the workings only later in their careers.

This book can help you get a job in the oil and gas industry, but it will do more than that – especially now that the world is thinking in the direction of renewable energy and the preservation of our environment. *A Career in the Petroleum Industry* looks at the need for the participation of recruits, not only in current oil and gas energy solutions, but also in human capital transition. As Nigeria contributes about 3.8% of carbon pollution to the world, we may consider ourselves as some of the players with a responsibility to the energy transition process. We have to consider the energy mix as a way of reducing pollution, while still upping our current energy capacity to provide the development needed for that transition. We need adequate healthcare and environmental good practices. For now we do not have the alternatives to solve problems of desertification in our own quarters, as is occurring in the Lake Chad basin. We become even more prone to crises.

A Missing Role

It seems that elsewhere, oil and gas exploitation is running concurrently with a quest for a new energy source, where the finances from oil proceeds are partly directed to research on renewable energy. What roles do the current oil workers from developing countries have to play in preserving the only habitable planet known to us? Will we suddenly find ourselves abandoned as the whole world, including expatriates in our own country, move on? Would we be able to man any leftovers of the billion dollar infrastructures already installed in our oil fields? The success of a new energy depends on the alignment of technically competent oil and gas professionals to take charge and become responsible independent energy resource persons.

Without a sense of direction, many of them will live on auto pilot, while the resource holding them like a platform collapses under their feet. It is good intention that leaves a legacy. We are all faced with challenges, but a noble cause and course make a river meander through rocks and thorns. A crooked river is different from a meandering river only by their difference in manoeuvre. There is a mindset that has developed over the years, making oil and gas elites in developing nations lose touch with reality regarding society, patriotism and common sense. Most of them remain skilled but superficial, crooked instead of noble.

This book will give you a glimpse of what goes on in the industry, gaining some clues from my own journey through an oil and gas career path over the years. During this time, I have seen that despite the supposedly good financial remunerations, the challenges of self fulfilment abound, both at home and abroad. I learnt some of the deepest unwritten guiding codes for a sustainable career in the oil and gas business. I found out that everything: recruitment, training, work culture, promotions and eventual systemic control of individuals, is driven by these codes. Once you realise this, it becomes easier to foresee and overcome every challenge that comes your way in this arduous, albeit rewarding, experience.

By the time you reach the end of this book, you should be able to connect and discuss with people in the oil industry. You will learn how to become involved and the role you need to play. You will know what lies ahead after the ups and downs the industry is currently going through. The aim is not to go into the technical details of oil and gas operations, but to get to know the in-house culture and trend. What has changed since they got that job? Let us follow the personnel and understand the roles of the hierarchy, as they climb the corporate ladder via promotions, the oil business and the system that runs the day.

The Turning Point

In 2004, I obtained a life-changing scholarship, thanks to TotalFinaElf (Operator of the Joint Venture with NNPC), now Total E&P Nigeria Limited, and the French Government (through its embassy in Nigeria). That year, there were two postgraduate opportunities to PhD level, after a selection process across Nigerian universities involving a merit exercise. Besides an oral interview, it involved aptitude, written professional, physiological and medical tests. It was not just a funding to study, but a closely followed up in-house development programme entitled: *Development of Talent Award.* We assumed others did know, but for the two of us selected for the programme that year, knowing exactly why the cultural, academic and technical immersion was so intense was a challenge. I remember asking why so much effort was being put in from different quarters to train us if there was no catch.

We had our own goals and we were happy to go abroad for our postgraduate studies. I was to study professional courses, but only after an immersion into the French language and several cultural programmes (including those with the European Socrates and Erasmus exchange programme students). I obtained my Masters and PhD, studied professional courses, trained at work, but each time getting to junctures where it was obvious there was something more to it. It turned out that the programme was a tug-of-war, tied by high executive tussles, pulling from both sides, between a few good national and international progressive persons in the industry and, on the other side, loyalists to a system of proxy for multinational oil caucus.

This scholarship and opportunity to mix with international free-minded persons, with radical opinions on the oblique relationships between the corporate headquarters of multinationals and their affiliates, and the training and work experience at technical headquarters in Europe, was the beginning of an intriguing journey whose aspects you will learn in this book. As mentioned above, it turned out that I was in the middle of two tussles, each pulling me to their sides: to use or to be used – to join or to check – what you will understand in simple terms as you flip through the pages.

As the only survivor of that year's *Development of Talent Award*, I express gratitude, especially to my mentor and his allies for showing me the light, and for trusting me to bear this burden of truth in turbulent waters. They took the struggle for the development of oil and gas host nations, for social responsibility, to the next level – showing that true transfer of control or mastery of a nation's natural resources, is not given or taken by force, but granted in style. There is a time when there is nothing else you can do, but to carry people to a visionary destination before they understand why. Where, as Miguel de Cervantes puts it, "Truth will rise above falsehood as oil above water."

Join me to experience another angle of the saga of the oil and gas struggle. It all starts with who is employed, the process and the objective. You too can become employed if you understand the expectations and present your application accordingly. This book outlines the process, the training and guides you need to get a job in the oil and gas industry and how you may play a vital role in this dispensation. The ball is in your court; it is time to see both sides of the career coin in the petroleum industry and then make a choice.

Chapter One

CAREER: TYPES AND ACCESSIBILITY

Oil has been a source of livelihood since time immemorial; it served for light and spun our wheels. When the world thought that petroleum was about to decline, unconventional hydrocarbon happened.

1.1 What do They Want?

The first thing to ask if you want to work or do business with an oil company is: What do they want from me? You may not have the skills now, but will you in the future? Are you teachable? Will you be loyal enough and focussed? These are the thoughts of the executives in the final panel meeting before your employment. They project you into the next five years and see what you will deliver. Dedication is crucial.

Today, loyalty has been bipolarised: many want to know what you can offer the major oil companies, but a few select persons want to know what you will offer your nation. The governing authorities are making this clearer and not hiding this reality on paper. They insist that every contract with foreign oil and gas companies must provide opportunities for Nigerians to obtain training. The IOCs must be willing to transfer technology over time. They want Nigerians in the industry to gain the knowledge, quickly. The regulating authorities want to have representatives and a supervisory seat in every oil and gas company in Nigeria. However, the oil workers in private companies will ever remain faithful to whoever pays them directly.

Everyone agrees on the need to transfer competence to local personnel. The speed with which it has to be done, that is the issue. Our foreign friends want it to be more gradual. Expatriates leading the major oil companies want to supervise the staff closely, especially those who might be highly skilled. They get to know everyone recruited; they predict how you will behave in the future. The future is very important because it determines the present, not necessarily the opposite. The struggle for control today is because of tomorrow.

The problem for the multinationals is that some recruits and few existing local elites in the industry want to be in charge and direct that future for independence. Understanding this will determine to some extent whether you will get in or not in the industry. Your approach when you appear for an interview speaks volumes. The people at that table are thinking about what you can offer, corporate or technical, and your loyalty. They want your skills or can help you develop them, but you must be most loyal to your recruiter. Your success in the Nigerian oil and gas industry in the near future is hanging in the balance.

Some oil and gas multinationals select a few for recruitment and carve out a niche that only that select few can attain, those who have been profiled to be malleable in the struggle for oil and gas control. The recruitment and career processes are all in line, focussing on benefits, a lone training and growth pathway and aspiration for a high position. Later in the recruit's career, what they want will evolve, and they begin to see how the choices made have taken them to the point of no return.

1.2 Permanent Staff versus Contractor Profile

There are two types of job opportunities: the permanent and the contract or temporary. *Permanent staff* are employed to become part of the company. Right from the start, they are given the impression that the company becomes theirs on recruitment. Not necessarily because this is true, but trust builds up and some permanent staff will then be in a position to supervise the contractors.

Contractors are engaged to work temporarily for the company. They work alongside and sometimes within the same offices as permanent staff. Some contractors work for years, even until retirement in an oil company, remaining contractors instead of full or permanent staff for 'good' reasons. There are two types of contractors:

(1) Those who work for other companies, but are sent by their firms to work for the oil company. They are seen as outsiders, made to work in a standalone system with limited access to sensitive data or technical information. A contractual term of confidentiality keeps them in check, even more limited both within and outside the company.

(2) Those who are individuals, employed as support personnel to work directly in an oil company. They are technically competent, sometimes more qualified than the permanent staff. They occupy positions that hardly evolve, may be paid less, even though they may have trained many of the permanent staff who rise to become their boss.

The question of who are employed as contractors or permanent staff lies in some subtle profile differences.

Persons recruited as permanent staff:

- Are young or started working elsewhere when young.

- Have had no break in their career and went from school to their first job, internship or training.
- Have a dependent mindset on the prospective employer.
- Show focus, loyalty and commitment right from the recruitment process.
- Score very highly in aptitude tests, but are not necessarily business minded.

- May come from another oil company with experience and good track record (which means they meet all the other conditions above).

Persons recruited as internal contractors:

- Have a wealth of technical experience and are competence on the job.
- May have had breaks in their career and not a straightforward schooling history.
- Have an independent mindset, and focus more on delivering the job than on company politics.
- Show less loyalty and are ready to work across borders with other companies or competitors.
- Do not necessarily have to score high in aptitude tests, but rather on-the-job competence.

Prior competence is not the determinant for employment as permanent staff in oil and gas companies, especially in developing oil host nations. High academic performance, even intelligence quotient, may be important, yet it is more of loyalty and an adaptable personal profile that is required. You can be trained to develop competence. The only time prior competence is absolutely important is in the engagement of consultants or contractors. Consultants are not treated as in-house employees, but ad-hoc service providers hired by the company for a given task. They are independent and remunerated in proportion to their specialised services.

In the long term, in-house staff are bridge builders between the company and their contractors, between the multinationals and the host nations. This turns out to be their major role. Unfortunately, that role has not been played patriotically. When a person with a contractor mindset is employed as a permanent in-house staff member, it could lead to what is known as 'recruitment error'. The contractor versus employee profile is one of the unwritten rules that control the state of affairs in the oil industry.

1.3 Recruitment Process

Recruitment is everything in the oil and gas industry. They must choose people of high calibre to meet the challenges of oil and gas development; it is becoming more challenging for reasons more than just a decline in resources and oil price. There is a challenge in determining which candidate to employ or not, which employee to keep or let go. The oil company is laying off personnel while also recruiting.

There is a crisis in the oil industry, where the new generation feels that the old is too adamant, while the older generation fears that the social media and online world may no longer allow for having tough and disciplined staff to man arduous oil work. The social media also opens people to the world, making them ask questions that were never discussed in times past. Individuals choose to have a social profile that may be completely in contrast with what they do in the real world. Sometimes these profiles create tension in the workplace.

The new generation expects participative management, where they can say what they feel and the leader adapts to accommodate them. The leader will draw strength from multiple ideas and allow the organisation to change and become more dynamic and interesting. The old leaders still live in the world where they communicate from top down. They are seen as boring. This is causing the employment process to evolve. The employment processes involve:

- Oral interview
- Written aptitude test
- Written technical test
- Assessment on task
- Panel interview
- Psychological test
- Medical test

Offers to experienced staff may take a shorter process than offers to fresh graduates requiring aptitude. While job competence is key for experienced staff, the academic performance (first class and second class, usually upper division) is a requirement for graduate intakes. For experienced candidates, you get a job offer before you resign a current contract. The new company may want to see your tax clearance to ascertain your previous pay, which they will then raise.

In standard recruitment processes of large oil companies, no one man can employ you; it is the process that secures the job for you. Often, people call and assume that a top executive can just offer them employment. That is possible in a small firm where one or two people make the decisions. Even then, the industry requires skilled solutions that no one wants to employ a friend or relation into a technical position. They prefer candidates that know what to expect, had been involved in some ways, and can deliver. Despite this, in some firms where politicking is the order of the day, nepotism is possible during recruitment.

1.4 Be an Insider

Someone read my book: *Overview of the Petroleum Industry* and called, lamenting that he has been seeking contracts in the industry for years without success. I suggested that he should first go and train, work or serve in those parastatals from which he wanted contract offers. I advised him to register in their association, participate in open Call For Tender (CFT) and network with potential tycoons to know what they wanted, give them the confidence and opportunity for them to know that he can deliver the job. He asked why he needed to do all this before expecting a contract offer. I said so that he will first become an insider before becoming a player. He was happy, yet still wishing there was a shorter way.

The industry works with their own. They work from within. You need to be an insider to get in. Due to the image consciousness of the sector, especially in developing countries where the role of locals in supporting the IOCs is crucial, the big employers ensure that your loyalty to the industry is ascertained before they allow you in fully. The job demands dedication. Have you been tested as an intern student? Have reports and evaluations been written about you during a secondment? Have you participated in a field trip where you met other explorers? Did you undergo three to six months of intensive training in the industry?

How do you become an insider? By getting closer to the industry – to the pools within, from which recruitment is completed. Look for scholarships and grants in your school early enough from oil companies or trainings and internships. Participate in sponsored meetings. Be as close as possible if you want to get inside. You are observed once you come close and you have higher chances when you talk and laugh like the others.

1.5 Application and Interview

Everything starts with the CV. The worst error is a *shadow CV* (without real experience, poor language, including not using industry terms, and time gaps showing several periods of your life not accounted for). Year to year, month to month, there has to be something covered in your CV life. Dormancy is a sign of free time, free mind and perhaps less commitment in the future.

You need to craft your *CV* or *resume* to fill up any spaces in your academic years or career, as your potential employer will be worried as to how you survived when you had no job, or appeared to spend some periods doing nothing. All through your career in the industry, you will be doing something. The short leave you will take in a year will be a busy time, when you meet up with multiple, outstanding house effects and solutions in your home. An idle man in the oil industry is as well considered a devil's workshop, and a free-minded person cannot be trusted.

If you are called for an interview, you may be asked to run through your CV, meaning you should orally recount your story by going through your CV step-by-step. They may make references to your CV, focussing on whatever stands out from the others. That is why in your CV, save time with the normal headers and personal details, which every other person will give. Your academic credentials should go in one line max for each qualification. They already know what it entails. The documents you attach are there for consultation. Thereby you create space to be identified when you add something unique. If anything stood out during your academic performance or you were top at something, then mention that.

There is a difference between the image of the oil company and the actual work on the ground. The real thing has name, places and jargon. Those in the industry will know when you talk to them. Those who apply for work with a CV based on the public image of the company, copied lines from web pages, will be isolated and their applications discarded. You can be forgiven if you are a fresh graduate. If you claim to have seven years experience (seven years period is often the beginning tag for opportunity of an *experienced position*), then you need to talk of rig, oil, logs or whatever else; you need to use oil well names and their locations to secure an operations job in the industry.

Even if you sail through the initial HR-based interview, you may miss it at the technical or panel interview. If you are a fresh graduate, learn the basics and language of the industry. Associate with the right people and place to get that tinge. Don't worry about perfecting the actual job skills. You will be taught just how to do that when you get into the company.

In my book: *Overview of the Petroleum Industry*, I gave a list of the oil and gas industries in Nigeria and elsewhere, grouping them from super majors, majors and national oil companies. Go to their websites and look for employment opportunities. Most of them provide information on what is required to qualify.

The Interview

Always say to yourself, it is real. I will do some actual work. Don't focus on the benefits, but on what you can offer. Once you introduce yourself as, for example, an experienced geophysicist with over seven years of experience, or as a just graduated geologist, move on to actual tasks. Say what you can do or would like to do (go to train and learn something before the interview).

An engineer seeking a *development position* will have to say things like this in an interview:

I worked at field X. We were the first to initiate the Y process in the Z platform. I was technical personnel for three years before taking up the role of pressure manifold supervisor...

Or as an internship student:

I was in the same office with Mr... who was my supervisor. We worked on Well-A in year YYYY and the field now produces.... barrels of oil per day.

Mentioning your supervisor is OK; it shows you have worked under supervision before. If asked what you can do, you may say something like:

I participated in the use of LWD to identify the reservoir interval in real time, which we correlated with X, Y, and Z in the offset wells before drilling exploration Well-B. We rightly predicted the PPP such that we could reach the target without issues.

Explain how you interpreted the petrophysical data, how the different logs were used. Add your contribution to the reservoir model for the oil well that proved....

Many potential candidates arrive with nothing at all to offer, with no information about the work they want to do. They say things such as: *I do high-quality work; I am dedicated and seriously committed. I promise to work hard on my duties....* Every other person is saying the same. It is not an emotional process; it is a logical one; unless you are to be employed in government relations or some beauty pageant receptionist role.

Get yourself familiarised with what happens in the industry, so you will be confident when faced with the players. They take note of those who know what goes on inside. You may visit the websites of the Nigerian Association of Petroleum Explorationists – NAPE (www.nape.ng); Nigerian Mining and Geosciences Society – NMGS (www.nmgs.org.ng), Department of Petroleum Resources – DPR (www.dpr.gov.ng) and Nigerian National Petroleum Corporation – NNPC (www.nnpcgroup.com) for more information on the industry. You may also visit other international organisations, like the Society of Petroleum Engineers – SPE (www.spe.org); American Association of Petroleum Geologists – AAPG (www.aapg.org); European Association of Geoscientists and Engineers – EAGE (www.eage.org) and Offshore Technology Conference – OTC (www.otcnet.org).

Register and participate in student chapter events and the annual conferences if you can; apply for scholarships, go to business meetings, most of which are listed in the above websites. If you are an accountant, HR candidate or marketer, these meetings will expose you and open opportunities to what is at the heart of the oil companies at that particular moment. The news section on www.iesog.com is available for continuous learning and opportunities, your gateway to petroleum and other energy resources.

Which company do you want? Even if you want to work in any one available, give all your strength to one at a time. Study and know that company, see yourself in there. One error I see in potential candidates is to be too liberal; they post mediocre content or hard radical activist information on their social network and even professional network account. Recruiters may check their profile. They type your name in Google and see what it gives.

Often, top management of serving staff google their personnel names to see what they are known for online. Versatile publications, even in your professional field, could give a wrong signal of someone who is too social to concentrate on in-house operational duties, unless you are being considered for a Research and Development (R&D) position. Are you rampantly applying for other jobs? Do you have other sources of income? They don't quite like that; they want to ensure that your workplace is your only farm land, your source of daily bread. They want you to form a dependency link. Have this in mind while crafting your CV.

Again, always highlight those places where you did exceptionally well. The oil company only takes people who are the best in one area or the other – it must not be making a good grade in school alone. It is also highlighting any project where you shined, so long as it is not in conflict with their goal or present you as a liberal. Oil companies do not take busy-bodies, at least not as full-time staff. I have seen some people who came for the interview and say things like:

> *I practice at home, offer my skill services to many other companies. I am also a business man, I supply goods and services and travel to China every month.*

Such credentials are isolated and may end up in the archives or in the waste bin right away. If you talk of yourself, let it be your development in terms of the job, not external entrepreneurship. Don't demand anything at the first stage of interview or application, ask what to give.

It cannot be over-emphasised that you should know the company you want to work with. What are their oil fields, wells and oil blocks? Why are they hiring you? Arrive in time. You may be given a relaxation suite or a night over, depending on the stage of the interview. Give attention to the secretary or anyone you meet in the office premises. They may be your future colleagues and besides, how you treat them speaks volume. Be organised; they probably are not going to take your nervousness seriously. It doesn't matter that much. Prepare for the most obvious questions, then be yourself and don't worry about panicking – a way to help you panic less.

At some point when they allow you to ponder or to drive the discussion a bit, or at the end of the interview, ask one or two reasonable questions. Let it not be a technical question with a specific answer, but a general question of curiosity or one showing your interest. Like asking what are the major challenges they face now or that you would be prepared to face should they hire you.

Chapter Two

EMPLOYMENT IN AN OIL AND GAS COMPANY

2.1 Career Opportunities

The oil industry employs candidates from every profession: engineers, geologists, accountants, doctors, marketers and more. It is a community that requires every service to run the infrastructure, governance and even continuing education. Some oil companies have mobile bank counters in their premises to facilitate financial transactions. Then there is the hospital with doctors and nurses, a maintenance unit, a network services unit, and a restaurant with cooks and catering staff. The core employees are the technical staff in the area of sciences, engineering, and human resources.

Employment may be direct (graduate candidates) or indirect (experienced positions). In *direct* employment, the company employs you as a fresh graduate and trains you, complementing the knowledge you acquired at school with work skills. In *indirect* employment, the company looks for persons already working, with the suitable skills or with reasonable long-term experience through internship, contracts, secondment or temporary attachment.

Oil companies and services firms advertise jobs on their websites. You also see oil companies' employment opportunities in industry bulletins and news magazines, including that of NAPE and other associations. Organisations such as Linkedin: www.linkedin.com and Rigzone: www.rigzone.com offer job opportunities in the oil and gas industry across several countries. Energy platforms that offer employment have different degrees of closeness with the industry. Examples of other websites include:
www.jobrapido.com
www.monstercrawler.com

www.info.com
jobcenternigeria.com
www.indeed.com
www.myjobmag.com
www.oilandgaspeople.com
www.hotnigerianjobs.com
ngcareers.com
jobsinnigeria.careers
www.oilandgasjobsearch.com
www.careers24.com.ng/jobs
www.careerjet.com.ng

Sometimes, all that these websites do is to lift an advert from the oil and gas firms' own literature and advertise it themselves. It is better to go directly to known oil companies or oil services firms and scroll to their career pages. Googling for jobs alone could present you with employment opportunities. Be careful not to pay to get a job or to physically visit an unknown destination in search of one. Find out about the organisation behind the platforms you see online, then seek a way to collaborate with them. You can start this while still in the university.

There are many kinds of collaboration which the oil companies have with universities and other citadels of learning. It has been observed that in countries where such collaborations are absent or abused, there is a disconnection between the students and the industry, leading to an absence of skilled candidates for employ-ment at graduation. When you send in your CV or if it is passed on to career scouts during a visit and it contains some participation in one of these collaborations, oil companies tend to single you out. They believe you have an idea of where you are going and have perhaps proven yourself already.

You need to participate in the various pre-employment stages by way of collaboration with oil companies. The collaboration with opportunities for employment discussed in my book, *Overview of the*

Petroleum Industry, include: University Chair, Joint Research Sponsorship, Professional Doctoral Positions, Internships and Industrial Training, Career Talks and Symposia, Professional Conferencing, Part-time Programmes, Secondment, Sabbatical Leave, and R&D. When necessary, experts in the oil company answer pertinent questions and solve problems relating to the industry by using standard research methods. This is why they seek out university students and lecturers to participate in these studies as part of R&D. You can be that student who is involved, even before recruitment.

2.2 Employment Avenues

By the time you get into the oil and gas industry, you already know what goes on inside. Even for direct recruitment from the university, the process acquaints you with what to expect. You see the professionalism, the rigidity, the wealth and the touch of oil. When you get in, you begin to see how that fits in with the work culture and why it was all important for them to get someone like you. Below are the avenues that serve as the final passage to employment.

1. Scholarships: Oil companies give merit-based or targeted scholarships to host communities. During this time, ensure you get in and start to work immediately or train with the company.

2. Professional (Social) Networks: Though you give out information on Linkedin, for example, the company searches for the one you did not want them to see on Facebook. As you network online, also network physically. Convert your contacts from the net to the real world.

3. Recruitment Firms: Note their limitations, as they only help to highlight that you exist somewhere as a candidate. Try to know the employer behind any offer showcased by a recruitment agency and know what that employer wants.

4. International Talent Hunt: When career managers travel abroad, they obtain names of potential employees from selected institutions. It is your duty to email, phone and contact persons in the company that they represent and follow up the process.

5. Graduate Recruitment Scheme: Whether done out of necessity or as a publicity/Corporate Social Responsibility (CSR) stunt, this is a good way to pass on your CV and do your best in the selection process if there is an interview opportunity. Out of 1000 persons addressed in an ongoing recruitment exercise, only about five may be employed immediately. But the door remains open for others who initiate communication, form and maintain the networking.

6. Youth Service Scheme (like Nigerian NYSC): Serving in a petroleum province, state or oil company brings you closer to employment.

7. Training and Workshops: Often paid for by an interested company, training and workshops, including field work, create relationships and introduce you to what matters to your potential oil company employer.

8. Referral (word of mouth): This works, especially for contractors and specialists. Keep good relationships and showcase your work. The internet is saturated, so also have real people talk about you. A written referral, a simple phone call or discussion at a meeting, will put a positive stamp on a candidate.

9. Employment Ads: Often targeted to direct graduates or indirect experienced workers. When seen in newspapers, it is perhaps too general and few will be selected compared to the number of applications. Follow the procedure, the CV and interview best practices, to get the best opportunity.

10. Job Opening: Experienced and proven persons, character, skills, and insiders, are the target. Often, this is a shorter process because, if say ExxonMobil is employing somebody from Shell, they already know his or her experience in the past few years.

11. Spontaneous CV Submission: This is the least of the opportunities on the employment ladder. Use it sparingly, especially if you are submitting to a few closely related companies. The oil and gas circle is small, so they may all be discussing you.

2.3 Work Culture

Everyone in this milieu is relatively intelligent. The recruitment process (the aptitude test, psychological test, the medicals, and more) ensures that. They are book smart, even street smart. Some persons who are new to the company attempt to show how savvy, witty and smart they are. Suddenly they find they are not alone; they begin to develop respect for old hands they had neglected before, especially when it comes to on-the-job competence.

You do not go around bragging to your company colleagues about your competence. If you do, everyone knows you are new. Only visitors talk that way. If you are an insider, you only talk about problems and their solutions. If people are interested, they will ask you about this solution of yours. If they don't, please keep quiet, because everyone knows it already.

It is just like the university, where after going through a tough entrance exam, you come in to find out how every other person did. The majority of students are as intelligent, if not more so, than you. In the oil company technical departments, personnel are trained on a regular basis and they do what they are trained to do, working for many years to rise to a senior position. So a Senior Geologist, Senior Reservoir Engineer, or Senior whatever, knows his *onions*.

It turns out that personnel, after the initial stage of assessing competence, are promoted for commitment and given managerial positions based on character (and not just because they are better at the job). Everyone would be better at the job at some time; the difference is in your interest and your ability to show it. Your employer likes it when you show that you are motivated. Those who seem incompetent were left out of the race or they chose to opt out. You may be moved around and lose the basics, or diverted to consolidate on a specialisation pathway.

Character as a requirement for a management position does not necessarily mean good character. Good in this case is subjective. Having an independent mind is not one of those qualities of good character in getting a management position. Loyalty and alignment to company objectives and following the working of the system, work for you. Not everyone will become managers – others may become specialists if they are good at what they do, but won't align fully to the ways of the system.

Be the doer, not the sayer. Occasionally you make presentations and propose a way forward only as a small conclusion. When you do a task, deliver the results of your work. When given a task, learn to say OK. This single word will save you a lot. It does not mean you accept everything, just means you act more than talk. Saying "OK" is better than "Sir".

On the rig, every personnel should know his job. Mistakes do happen, and are taken seriously. Mistakes are said to be good for learning, but not on an oil rig. You have to go through the checklist like a pilot and get it right or else there are consequences. A blowout on the rig (sudden gush out of high pressure fluid) leading to an explosion is as bad as a plane crash, if not worse. So when your supervisor gives a clear instruction, just say OK, then think and work on it and get back to him only if you think you need to make a change.

For once, you are among people whom you can reason with. Most are principled and smart. The only problem is that you get so used to such a logical environment that you begin to loathe the general masses, the street life, and some of your old friends that would have been a fallback in case you leave this calibre of colleagues some day. It will be foolish to expect your spouse to be that technical on family matters. You become less smart and the true meaning of intelligence begins to evade you.

2.4 Presence and Timesheet

The timesheet is used to control expenditure, recording who works at which time, how many days, and on what project. It helps to apply budgets, including salaries. In its budget role, the timesheet is rather for deciding which costs go to which department and not who came to work today or at what time. Emphasis is on an objective and achieving it. No one is monitoring whether you sat at your desk all day or whether you went to the coffee room every hour. Except when you are a new staff member with a supervisor just beside you, with your computer screen facing him.

In any case, you will be present, because most of the big offices are configured in such a way that you badge in the morning to the work campus. You are provided with lunch and other things that would have taken you out in the day (some oil companies also have cooperatives and shops in the office premises). You are in the office at the right time, even if not directly monitored. Once an erring member of staff is flagged, the curious manager and General Manager's (GM) eye could be scanning who spent more or less time at his or her desk. They pass each morning informally to say hello and ask the whereabouts of Mr A or Mrs B. While you badge in and out, the system automatically tracks personnel arrival and departure; this tracking can be consulted if and when necessary. It will not be necessary if the outlook on the personnel was good and nothing appeared to be wrong.

As for the hierarchy's opinion of you, it will be the attitude of too good or too bad. Your first appraisal will matter – if it is bad, then you have made a bad beginning and must work hard to make it change quickly. Others assessing you will take a leap (upwards or downwards) from that base. Again, it is like the first year at university. It's very important you get it right.

2.5 The Concept of Selling

It may take a long time after your employment, but sooner or later you learn the concept of *selling, strategy* and *communication*. This is a very important approach in the oil and gas industry, especially in affiliate offices implanted in developing oil host nations. 'Communication' in the sense of what do we want them to believe. You hear a senior staff ask the boss, or the boss ask the top management "What do we want to communicate?" even before starting a technical study. It is part of the team spirit, the work culture, the way it goes, to understand what is the central message intended for a given purpose. It is not about right or wrong, or about the technicality of a scientific quest. Yes, the initial studies may tell us where oil is and where it is not in the subsurface. After that, the question of how many oil wells are to be drilled, how much the cost, and how this is settled with the partner and authority is about what you want to 'sell'.

The *operator company* does the expenditures and bills the partner companies and the host national company. The operator convinces the regulating authorities about their actions. It takes a lot for the authorities to know the right questions to ask, let alone be able to enforce the answers they may never get.

There is a language of selling, which if you do not use, makes it obvious you are not in the game; you may be ignored or deceived during the meetings. One such language is 'strategy'. If you ask what the strategy is, it means: where are we approaching the 'communication' from? To go or not to go, to drill or not to drill, to say yes or no: to "make or to kill" the proposal. If the strategy is positive, then we work towards it. If not, then the aim is to 'kill' the proposal. You kill or make a proposal.

If the operator wants to drill an oil well which is not important, apart from the advantage of recovering expenditure and service costs earlier, then it may be said that the operator is 'accelerating' the production of the reserves. The new oil well will only pull more oil that would have been produced by other wells anyway. If the well is truly necessary, then it is said to bring in 'incremental' volumes or reserve. The reserve may not be increasing; in which case incremental reserve means the well is serving the real purpose of pulling in oil that may not have been produced without this additional well.

Success may be geological, economical or political (the word 'political' is avoided and replaced by "for administrative purposes"). An *appraisal well* may only be needed for *calibration*, to know where the stratigraphic levels or horizons, depths, and fluid types will be in the field. Finding oil may not be the objective. In fact, you may prefer not to drill at the point where you have the maximum chance of finding oil. You may drill an oil pool from the flank or side, so that you only find oil if the reservoir is large enough.

From studies, an operator may suggest that there is a disconnection between the current oil field and the nearby field, placing an uncertainty on the presence of petroleum in the new block, to ease the acquisition of the nearby field from the authorities. The young engineer or geologist who is doing his work at the corner will be so focussed on technical delivery, without knowing why top management is driving the result to a given conclusion.

When top management invites you for a coffee, then tells you that the proposal has to 'fly', they mean your technical result should show it works. You are only there to justify that conclusion. You are working from the conclusion to the question. Of course they look at the initial work and sometimes request an in-house objective work to find out what is the truth. The truth is different from the 'message' that they want to 'communicate'. When you are told that the aim is to "bring them onboard", it means you are to convince the others of the message you carry and not attempt to reason out other possibilities. You are to sell at all costs. When a meeting is called with partners for alignment, the aim is to get their support before meeting with the authorities or the regulators. An "alignment" is an informal agreement to be formalised in the official meeting.

Imagine you do not understand the details, the language, the technical points, the diagram and the processes. You are deceived, sold alongside the message. Many local staff are using this language, sometimes selling their birthright, their resources and their integrity, without realising what they are doing. They have learnt it over time and become sensitised on being strategically correct, to the detriment of their innermost ethical values.

To go against this language and attempt to be too straight means you do not understand the working of the system, its interest embedded in the oil-producer to host-nation relationship for so long. You are getting in the way of management, your boss, and 'common sense', and these may affect your appraisal as not being good in 'communication'.

2.6 Role of Appraisal

An appraisal is a routine assessment formalised to keep a staff member on track, by giving him feedback on his performance and deciding on the next step in his career path. It involves an interview where the achievements of past objectives are scored while setting new ones. The appraisal meeting is an important process in the career of a staff member who may also be given the opportunity to give his own feedback on the job. Evolution of promotions, remuneration and benefits are tied to appraisal. It turns out that in most cases, the appraisal is perceived as biased. This is because:

- The unwritten feedback, communicated to your higher hierarchy by your boss, counts.
- The smaller lines may include comments that call for worry, even if the larger scores say all is well.
- Your scores might have already been decided before the interview, whereas you think you are arguing a case.
- It has become a tool in the hands of managers or top hierarchy to subjugate employees to a given way of work execution and behavioural attitudes.
- Lower level consolidation, whereby limited benefits are available for your department, result in low scores, despite high individual performance.
- The appraisal result may not be tied to actual rewards, where scores are allotted just for face value.
- The unique long-term development path already planned for a staff member may mean that the appraisal for such personnel is only a formality.

The appraisal method in most oil firms has been so contested that a lot of techniques are employed to ensure it is *time bound* and *measureable.* Despite that, it remains a tool in the principle of auto-motivation and focus. The result is influenced by an opinion about you, even before your work performance. If you have multiple interests and sources of income, you do not have the right profile to be entrusted with responsibility within. You are profiled out, and when you see your colleagues being put over to manage you, you now know why. In your end-of-year appraisal, you will see something like: "You need to be more focussed." That is, for people who tend to have other investments or business while working with the company. It is a warning appraisal note with placement consequences. The company may decide to send you on an intensive indoor training programme in another region or country for months, or to an offshore base, or somewhere from where you cannot coordinate any other business.

Another such warning comment in your appraisal is: "You need to improve communication with your hierarchy." Improving communication may mean you have to listen beyond the words, see the trends and align to top management's actions, whether you consider them good or bad. You sell whatever product is placed before you.

2.7 Job Profile and Hierarchy

In this professional setting, no one is king. Focus is on the common goal. Adherence to company policy is key. There is this shadow that will always hang over you, a big something that you do not fully understand. You keep thinking you will get it when you get up there, but no... it shifts away. It is the knowledge that the stakes are high, that there is a crushing risk at all times. Society is so much in need of energy on a daily basis that the authorities just have to let go.

Each staff member will attempt to align to the business. If oil is produced and profits made, the trend continues and everyone is happy. Eye service is not needed. Your boss may laugh, drink coffee, chat and go partying with you. But you must know where to draw the line. Of importance is quiet loyalty – which you show by supporting his vision. Don't challenge that project; offer your advice as if asking a question or considering a point. Don't relegate him in public or in meetings. Once this is clear, the eye service is less important.

At the early stage of their career in an oil company, this absence of eye service makes personnel think the hierarchy is weak. The young recruits learn to work and only report to a manager who emphasises team work and is better known just by his name. Later, they find that they are individual employees with specific expectations.

The role of hierarchy starts to weigh in more and the staff become more sensitive as they begin to understand their responsibilities. Few people reach the very senior levels where peer competition takes a different turn, silent but dicey. The stakes in the job demands that you keep to the rules; a shadow, which has more impact on your day than your real boss, will keep you under control.

2.8 Career Areas in Unconventional Resources

Most existing opportunities in Nigeria are in the conventional hydrocarbon resources. Conventional petroleum resources refer to *crude oil, natural gas* and *condensates* that are less viscous and stored in porous sandstone or similar permeable rocks known as reservoirs. When drilled, the oil and gas flow into the well from where it can be pumped up to the surface. However, unconventional resources provide untapped opportunities for many in oil-producing provinces.

Unconventional hydrocarbon is petroleum produced through other methods, where drilling alone is not sufficient to cause oil and gas to flow into the well with minimal effort. Unconventional hydrocarbons include *liquefied coal,* biomass or plant-based *biofuel,* and any reaction of carbon and hydrogen yielding substances to produce petroleum products. It also includes *natural bitumen,* and fossil fuel which occurs in *oil sand* and in *shale rock* with a high organic content.

Working and building a career in unconventional petroleum resources present several new opportunities. Unlike conventional or traditional oil and gas production where workers must go through the pains of confirming a source rock, the reservoir rock and the trapping process that seals it, an unconventional resource is much more direct. The *source rock* for generating petroleum is sometimes the deposit of interest in the production of petrol, irrespective of trap, seal or timing of oil migration, which would have been important if it were conventional petroleum.

In Nigeria, unconventional opportunities are just emerging and many are still unaware of this. Tar sands occur in high concentrations in Edo State and in lower concentrations in both Ondo State and Abia States. Coal occurs in many states of Nigeria, dominating the south, south-east and all the eastern and north-eastern axes, including Enugu, Benue, Kwara, Anambra, Ebonyi, Abia, Delta, Nassarawa, Plateau, Bauchi, Gombe, Adamawa, Borno, Yobe and Sokoto States. The potential of oil shales in the Anambra Basin is being re-evaluated. Gas resources that until now have been ignored, some of which are trapped in the coal beds, may turn out to be the main source of electrical power for these states in the not too distant future.

A new workforce is needed and staff training will need to be adapted to the changing scope and environmental challenges of unconventional prospecting. A mix of experts and basic hands are needed in the career make-up for unconventional resources. Everyone has a role, from the health, safety and environment (HSE) personnel who do the risk assessment of environmental hazards, to the site geologist who does the prediction of resources and assesses deposit quality.

The first step in prospecting for unconventional resources is to have a geological framework and an understanding of the basic natural petroleum system. The initial investigative work onsite involves answering basic questions, such as the stage of maturity of the unconventional deposit. As production starts, there is continuous feedback and post-mortem to compare predicted deposit size and actual results. Horizontal drilling, possibly fracking and unconventional production in Nigeria require an adapted workforce, and that is one of the solutions that innovative oil service firms need to provide. Areas of training, career opportunities and services in unconventional resources provided by the Integrated Elvee Services Limited (www.iesog.com) are discussed below.

Biomass and Biofuel

Personnel working to produce fuel from biomass would have to use plant and animal materials, organic matter, to produce liquid hydrocarbon. Biomass processing centres function as biorefineries, using farm and animal waste to produce fuel on small and large scales, thereby employing a large workforce. Skills needed to convert animal dung into biofuel already exist in traditional methods, as well as advanced technology. It is profitable when meant for local use, where the biomass is converted to oil for nearby utilities. There would be less pollution due to decomposable organic matter. Assessing the relative benefits of generated fuel, from large economic conversion of non-waste cassava chips to ethanol fuel, compared to direct use of cassava as a staple food is an important consideration. The problem with plant-based fuel is the depletion of biomass in nature. Using waste and dung makes it rather a positive derivative of second value.

Oil Sands, Tight Oil and Oil Shale

How many oil fields in Nigeria have been abandoned because they were considered too tight, with less permeability to produce enough oil for major oil corporation? *Tight oil* and *oil sand* would become viable if unconventional methods such as fracking, acidification or hot water injection were adopted. This is especially true for smaller-scale production, for low stake use in neighbouring communities but not for the larger international quota of production by the IOCs. Marginal oil and gas fields could be reactivated by smaller indigenous firms or their life extended by unconventional solutions.

Thermal Depolymerisation

Thermal depolymerisation is a process that converts waste material with hydrocarbon potential, which when heated, generates oil and gas. Car tyres and other petroliferous wastes can be artificially heated to produce petroleum. It is not the most favourable process for supplementing the generation of hydrocarbons, due to the ratio of input to output hydrocarbon volumes, but other by-products of waste tyres, such as shoe soles, may make it worthwhile in local communities. The skill is one that could be acquired by willing persons.

Coal Liquefaction

Coal liquefaction by a process known as Coal-To-Liquid (CTL) converts coal to petroleum products. The techniques are old, with new advancements, which is why it requires both local blue collar and skilled personnel. It is a major source of employment in countries like China and South Africa. Nigeria has abundant coal deposits, which is why it is a major consideration for creating employment and powering local energy grids. One of the challenges of coal exploitation is excessive use of water, which may impact on communal rivers. There is the need to abate pollution by best HSE practices. I personally recommend only a minimal exploitation for local use.

In retrospect, why hasn't coal liquefaction to petrol been considered in the coal era in Nigeria, even before the recent concerns of the impact of fossil fuels on the environment became exacerbated? The conversion process, even for the production of briquettes, which are compressed blocks of coal powder for use as cooking fuel, was doable and easier than most technologies already mastered in the country. By ignoring some of these basics, we lost grip on continuing technological development for an independent energy base. Maybe this is the time to act. There is a new opportunity for unconventional resources in sustaining the job security of oil and gas workers in this era of energy challenge.

Let us not make the mistake of the past. Let us not only understand the techniques of a new energy quest, but the need to start from the basics and aim for energy independence. A country like Nigeria needs to explore the career paths that determine the role of oil and gas elites in the petroleum industry. The reasonable efforts to properly manage the upstream sector have failed, which errors persons pursuing a career in the sector must mitigate to succeed in this new quest for diversification of resources.

Chapter Three

JOB SECURITY AND WORK PRESSURE

3.1 Lay-off and Replacement

The oil companies keep the personnel they trust at all costs, since they cannot be sure a replacement will be as faithful. Job security in the oil and gas industry depends upon whether you work as a member of the main cooperate body, or a service hand in a routine job any other person could do. Contractors are easily laid off in cases of redundancy. For permanent staff, the company has invested in them and only in specific challenging situations will they want any of them to go. Permanent staff members have:

- received long-term training
- been immersed into a belonging culture
- been psyched to be happy only within
- a risk of badmouthing when they leave

Among permanent staff members, there is a hierarchy which shows increasing job security:

1. Metier-oriented personnel (geosciences and reservoir, engineering/drilling personnel) will be retained while the big project lasts.
2. Person-oriented personnel (cooperate business managers, human resources management, key interface personnel) whose jobs are based on relationships and holding the base together.

While contractors or service hands are very useful, until now they are unfortunately treated as if no longer useful during periods of restructuring; as if there is nothing to gain or lose by their getting in or getting out.

Something new is happening in the industry. Multinational oil companies are employing temporary service hands and contractors to man major positions. It seems this provides the multinationals flexibility, less interference from a strong labour union and the ability to run the affiliate countries from foreign headquarters. Independent contractors and service contract staff are not members of PENGASSAN – *Petroleum & Natural Gas Senior Staff Association of Nigeria.*

Such personnel have less control, less ability to be independent and must hold tightly onto their jobs due to weak contracts. They cannot hold the company to account in matters relating to government policies and regulation, which becomes more imperative at a time of divesting and restructuring. These are signals that the multinational oil companies are re-strategising, attempting a single worldwide management system as a way of tracking the final stages of exploration and production in the affiliate countries. This increases their capacity to stay, leave or abandon.

3.2 Hard-to-Get In and Hard-to-Get Out

If you are in the big oil and gas circles, what will become of your career experience is the result of an existing complex oil and gas system of business. If it seems, from the interview process, that you will not fit into this organisational structure, then your chances of employment are slim. If you slip through recruitment and adapt to the workplace, then it will be difficult for the company to let you out.

Schemes and benefits keep you well within a circumscribed circle that is insulated from society by an invisible wall. These may be loan schemes, internal loans, and external loan reference letters that authorise you to obtain more loans from commercial banks to buy life's comforts. They could be car grants, housing grants, or land grants that amortise over several years. It could be free lunch, escort car or home and office security. Also on the list are post-retirement benefits such as health care.

In a tight economy, we see that this health coverage is limiting staff to company clinics where, as oil prices shrink, the available medication shrinks. The risk of kangaroo medical care for staff increases in developing oil host nations, a problem many of the labour unions cannot see. While in active work, some oil workers take their children to look for solutions outside and pay to ensure proper health care.

Beyond these material gains are the emotional grips and work ranking, positions and events that gradually define your life among birds of the same feather. With comfort and indiscernible cash heist, and virtual stability, comes emotional tension due to the requirement to work even harder. You keep refinancing your loans as your needs grow and your wants expand; they are hard to amortise over your lifetime. You are now well settled into your daily routine of work and monthly pay; but you lack the time to enjoy your comforts.

As for the company, they do their best to ensure useful employees are retained. Even vacations may be hindered just to ensure personnel do not have time to build another life outside work. In our time, any possessive attitude of the company further drives personnel to wish for a less constraining work life. For many, it only remains a wish, allowing dissatisfaction to creep into their attitude to work. This creates a two-way instability in the career and personal life of oil and gas industry workers; something we are beginning to see more recently. As P.R. Rose once said, "Most highly risk-averse people (and firms) never realise the high price they pay for their conservatism."

If it turns out that the future does not maintain the former comfort the oil workers signed up for, many will feel lonely in retirement. Some die within a few weeks or months after resigning, perhaps due to unpreparedness for their later life, a disconnection with society and loved ones, a sudden break from routine, financial tension and depression. Don't forget that the profile recruited are those who since childhood have never had more than a month's break – since their primary school through secondary education to university. They get a job on graduation and have been working until retirement.

This is why I hold seminars on survival for oil workers in a declining industry, in an attempt to carry them along during the ongoing energy transition. New workers should be trained on knowing this as they get in, to have a wider scope, achieve more in a short time and retire happy. I teach people to do oil work or business and have a farm on the side. I teach them to see the good in the spoils, that the current challenge in the oil and gas industry is as much an opportunity as it is a problem.

3.3 Come-in and Learn-later

At the point of seeking employment, many were just seeking a means of a livelihood. Still with the freedom mindset from school, when they were free between lectures and had their own way. Once you land the job, without knowing it, you may begin to lose this basic freedom of heart. Some oil companies attempt to give their staff social avenues, freedom of expression, possibility of contacts with the public and many more conditions that make working seem more like daily life. The higher the stakes, the higher the pay and comfort, yet the more there is a closed work outfit for the employee.

Oil companies are supposed to be one of the best organisations to work for in the world; they pay good salaries and provide amenities for their staff. In the big multinationals, an office oil worker recruited to be an in-house staff member will:

- Undergo training
- Travel the world
- Then settle to work

No matter what your skills, there is always something fresh to learn and a new experience to gain. Your employers are patient; they keep you involved while time passes and while you master the job. Over time, the personnel gain enough skills, become senior and able to work on the job independently. Then comes the next stage, where your work takes all your life – you are called in day, morning and night. You could be anywhere at any time and there is so much to do that you sometimes wonder how you got so tied-in. You gain job autonomy and lose autonomy in life.

In this information and data era, everyone wants more autonomy. It turns out that over 90% of the people in the big companies are now working for the salary, no longer because they love their job and have a sense of fulfilment of their long-term professional dreams. First the dream was to get the job, then to be good at the job, then to be free from the job. By then, you have severed links with old friends and relations; you hang in there hoping things will get better. It gets worse in most cases. Only a few recruits plan with their later career days in mind.

3.4 Comfort until you Come Forth

Comfort: Despite armouries of comfort around them, most employees in big oil companies do not enjoy their wealth to the full. Perhaps, some of their relationships do, yet not their closer ones such as with wife and children, who spend less and less time with the oil worker. It's the kind of wealth that carries a burden.

Oil companies invest a lot of funds into logistics, human and operational costs, in such a way that everyone is comfortable at first. Young members of staff are happy with the hotels, the luxury. Soon, they discover that comfort is not always enjoyable. You may stay in a 5-star hotel for training and only have an hour to use the facilities, because you are busy with the seminar and late night dinners. You wake up by 5.00 a.m. for the training programme and return to your 5-star hotel room by midnight. It still looks good, but many other things you appreciated before are whizzing past you. You are losing control and gliding into autopilot mode.

Higher life: An employee is suddenly immersed in a life style that he or she has not yet attained. Because you are placed in the topmost hotel during a work travel mission does not mean you are rich. Could you afford that on your own? Personnel unwittingly yield to the carry-over effect on their personal budgeting. They try to maintain such a life even when not in the care of the company. This means staying in the quarters of the rich in their early career years, where they pay very high charges. Such employees may never experience true wealth if they buy houses in the same neighbourhood, with credits whose monthly interest is above their salary.

Forth: As you become a senior staff member, you begin to see that this is not just comfort – you have responsibilities. If you have a family, children going to school, they may invite you to an inter-house sports competition where your child would win a medal. However, you need to be in the unending work meetings irrespective of your mood, your day – just like a soldier, you are called up. Most staff at this time learn that they are actually useful, not just having good time in the company. You have gone too far to go back, and forward you go. There are no free gifts in the oil and gas industry. Interests win.

If only you make a choice early in your career – know what really makes you happy and then stand for it. For some, it is working daily and climbing the hierarchical structure. For others, it is building a real estate. For others still, it is becoming the best oil rigger in the world. Follow your heart, no matter the consequences; work on the job while it lasts within your life plan, without letting excesses, such as overlabouring for promotions and material gains, ruin your goals. Then you will find much more satisfaction in old age.

3.5 Peer Pressure and Control

Peer pressure is probably the strongest controlling force that keeps people competing in the office, sometimes to the detriment of their greatest innate dreams in life. It is not the salary alone, or the pressure from hierarchy, that make staff forget the initial goal they had when they first became employed. They carry on the competition in school to the workplace, competing with peers. At best, doing what every other person did to remain ahead. When it comes to peers, they are your day-to-day companions that you cannot disappoint; your ego is high.

Oil industry peers are the hardest to beat, like in most other high tech jobs. They are as intelligent and competitive as you. Your success is not appreciated unless you belong to the cliché. There are inner circles and sub-groups among the peers. Within one circle, everything that happens is interpreted in a certain way. If a colleague in one circle resigns, they will say he has got a better opportunity. If a colleague in another circle resigns, they will say he has been laid off.

Employees often find out that one of their peers who was recruited at the same time as themselves has now become their manager. Some may not care at first, until their new manager makes one or two comments, or other people tell them, "Oh dear, your employment set is now your boss?" If you have a long-term career plan, always tell yourself that this is a passage. Later, you may find that you and your hierarchies are not that far different. Everyone is running to meet life challenges and no one really cares about what choices you made.

Maybe your dream was to work for five years and save enough money to move ahead in life, or just to be the best reservoir engineer in the world. You find yourself now working late hours to impress a non-interested boss, down-stepping on your colleagues, to put yourself forward and be noticed. Maybe your dream was to live with your spouse, but you are now working months away from home, ignoring quality time with your family, due to work load. It sometimes turns out that your personal life was better before you got that management position.

If nothing else, a regressive boss pitches colleagues against each other. Once they are in the competition, then the floor is set for other jeopardy. This is cruder in developing countries, or circles with extended job and family relations. In one case, the wives of the colleagues had their own circles where they met as friends, and one wife asked the other, "if your husband (team member) has finished the task which my husband (boss) gave him." It was a touchy one because these two husbands got into the company at the same time and cared about job positions. Few people achieve their life dreams because many let other interference, temporal disappointment and peer pressure, change their own life objectives.

If not for the peer pressure, many oil company office challenges would not matter. Peer reasoning and comparison are installed early in one's career, when group recruitment and the induction process highlight different career growth possibilities. This sets off a competition among recruitment sets and incites an internal struggle for one to be better than the other. Hence, everything such as transfer, promotion, reprimand or achievement for one employee matters so much to the others, irrespective of overall good comparative welfare and working condition for everyone.

In my own career experience, I learnt to focus on the overall picture. People came close to me when they thought my career was skyrocketing and became distanced from me when they thought my career was in jeopardy. Yet my own life dreams controlled my actions and my visions remained a guide; they helped me to know when to give up temporal gains for a long-term plan. I knew that if you do not have a target while you shoot, your missile will be carried by every wind.

3.6 Recruitment Error (RE)

Due to the long process of scrutiny during recruitment in the oil industry, it is rare to employ someone who is completely unsuitable for their intended duties. Most of the employees have basic IQ and show availability and motivation as they begin to work. For the few exceptions who may not, long-term training, coaching, inculcation of objective and work culture will make amendments. When all these fail, it leads to what is known as recruitment error. Recruitment error results when a member of staff shows:

- no motivation
- overly complains
- is not good or improving on the job
- does not find a place among the team
- does not buy into the policy of the company
- is not economic in the long term
- has an activism mentality

A sign of internal activism, like challenges from the company branch of the labour union, is not an issue – it even helps the company to feel good when they comply with demands. Activism within the union is a career strategy for both parties, just like opposition in politics – all are politicians. The union raises dicey issues before the annual negotiation of salary increase and benefits and threaten the company, then the issues die out after and resumes again same time next year. What is not desirable by the system is activism outside the scope of the company union, like going to press. Activism in terms of societal rights causes quibbles for your employer in the industry.

The industry avoids laying off quickly, rather they learn to deal with recruitment errors. A few causes of staff demotivation can be taken care of, such as trying to integrate family challenges in planning staff position and placement. Games and clubs are made available for staff, not just for their physical fitness, but as an incentive to have fun and forget their sorrows. Oil company career managers engage staff in several consultations to know where they expect to be in the next three to five years, in the short and medium term. They have the best career managers in the world, who attempt to align the staff ideals as close as possible to the company objectives. However, there are a few scenarios that make this impossible.

One is where inability to adapt to the working environment and people is related to physical traits and tics. A physically handicapped person could be well integrated into the system and everyone understands. An apparently normal individual who spits on the desk, washes his hand in a drinking cup, sends out a foul smell during meetings, fights with every staff member and especially with every boss, may create an atmosphere of resentment around him. Eventually such behaviour could get in the way of personal as well as formal work relationships.

When you don't fit in, it shows. A protracted depressing mood may lead to transfers and changes in position. Eventually, the situation may be regretted by the employer and considered a recruitment error. Such a member of staff is left to wear away in some inappropriate position, while allowing time to find a suitable exit. Those challenging moments can be quite decisive for the personnel. Some may recover and rise up to greater heights, while others fall into an irresolvable trench in their career. All that is needed is a twist, a small trigger that makes people come to a point of realisation that changes them forever, for better or for worse.

Often few people in an oil company are at the edge, on the balance, tending towards recruitment error. Some examples are:

- *A staff member who was abused by a former boss, who has lost the original spark with which he or she came into the industry.* Probably due to unchecked abuse, the employee has resigned to fate and no longer believes in the organisation. Such people may be rekindled by some push, a transfer or promotion. When not, they are better out of the way, quitting or a lay off may work for both parties and the company often comes to that realisation first.

- *A staff member who was groomed for a role but discovered he was just not made for such a task.* As an outside example, in some countries, people are groomed to become suicide bombers and may grow to discover they should not be. The oil industry is known to sponsor certain people to study in their headquarters, meant to eat and drink and think like the masters. There is always the challenge of whether such an individual will remain loyal. Those who attain a state of realisation may turn their alignment to their fatherland, especially in these days of increased knowledge and wisdom.

- *Professors and academia who may be too 'learned' or university inclined, before employment into an oil company.* Although this happens rarely, many of them would want to be employed, especially in developing countries where the financial incentives of the industry far outweigh that of the university.

Have you wondered why companies don't employ the best university lecturers in the world, to let them practice what they know best and have taught in school? The simple truth is that they can no longer be moulded to fit into the company's working conditions. When some oil companies, on rare occasions, employ such academia, they subject them to several trials and take far too long to grow trust in them. If you were a university don, now employed, there is a feeling that you have another plan, that you are not challenged enough – often interpreted that you are now too aloof to do the work.

- *Highly distracted people due to discrepancy in interest.* The company needs well dedicated people. For example, with a PhD, you may become a top highly smart executive or you could become a recruitment error – if you don't manage your high educational levels or if you are not well managed. The same thing applies to employing highly talented persons in the arts, like music. You cannot, as it may in your employer's eye, be a top musician and a top driller.

One is considered a recruitment error if attempts to reinstate confidence have failed. Sometimes this incongruency between the candidate's interests and the company's objectives were noticed during the recruitment process but ignored, either due to politics or quota system, or other employment obligations. The person has passed through tests, physiological examination, an initial three-to-six-months' trial period trying to adapt, yet what was ignored becomes obvious later.

The major oil companies take care not to release the strings of self-restraint to possible disciplinary actions too quickly, while the staff member remains an employee. This leads to an ever-increasing tension between some staff and the hierarchy. People who have been victimised and denigrated to a point of dismay in a company may be designated as RE, but are rather victims of the system. Being treated as an RE for so long can lead to a general feeling of abandonment and depression; some victims are unable to fully function elsewhere when given another chance.

3.7 Square Peg in a Round Hole

You might well have been a good employee, but something went wrong in your placement. Whether work position, role or location, each may have an attachment to staff core desires. It turns out that if you teach a fish to climb a tree, the fish struggles until death. For a while, changing a staff placement will not matter, but soon the reality dawns on the struggler. If you have been set up badly, where you do not belong, chances are that you fag out and prove the opposite of what you are. Over time, it affects your character. Character is everything in the industry. Yet personnel character is being tampered with all the more in this time of energy transition, cost-cutting, staff optimisation and position swapping.

These swaps in assignment are explained in soft tones, presented as multi-tasking or cross-functionality and part of the general scope of your career development. Have you seen a specialist reservoir engineer made to work as a drill-cuttings sample catcher, a sedimentologist made to work as a rig floor supervisor, an operations geologist made to file seismic data as a store keeper? These changes occur when personnel are moved to a new role in a new team, when they are under much more pressure to perform. If you show signs of weakness, your new entourage feels it. It gets worse when you notice and react; a positive reaction delays the frustration, a negative reaction sucks. Over time, a square peg in a round hole begins to look around or begins to refuse to fit. Then they are judged as being stubborn. If you decide to accept the situation, then you look stupid. A staff member bearing pain for a time may lead to a sudden gush of anger or an even worse reaction.

The new team members may be nice, kind and happy with one another, so if they judge your reaction as being difficult, everyone will believe them. Your own self confidence begins to wane. You lose what you need to overcome all challenges. At such low morale, you can be bought at a low price. A top hierarchy may propose anything and you will accept it. You find yourself working against your purposes, keeping silent in the midst of injustice or high-level manipulation of truth against your people, your country or your society. The comforts you have earned, the oil money, the financial rewards, come at a price.

Some persons are strong when facing frustration over long periods of time. They brave it, hoping that people will change, that times and interests will evolve, that the stakes will overturn. True as that may be, those external changes take the same time as the internal changes in you. The longer it takes for things to change around you, the more you would have adapted to remain unchanged by them. One good piece of advice at this time is to be open and speak out to the hearing of others, being honest from day one. Unfortunately, many have presented only the good side of the coin for so long that they cannot speak out about their challenges in the oil company where they work. Speaking out opens your mind and heart to new solutions outside the box, outside the sphere of the challenges.

What will keep you going in tough times is your ultimate dream. Do you have a second plan? Are you working to be free some day? If ultimately you want to make a career in a given company, then you need to treat some challenges as passing storms. Most of the top executives were once in your shoes. They were toughened, understood the system, only after there had been attempts to run them down. It is not about you, it is the organisation of things that prioritises the business sometimes more than the people. What you are going through has happened to others before and will yet again.

There are cases where an employee has been placed where they do not belong, as a kind of punishment for not aligning fully to the company objectives touching on patriotic responsibilities. When you earn such a high pay cheque from the company, some people say you have sold your soul. Your employers know how desperate you could be; they know that there are few organisations in the world that can pay as much wages as they do. They know that at this time, you are overly-dependent on the money. So long as you play by the rules, they guard you jealously, but when you stray, they press the button.

There are extreme cases where a staff member is in distress and needs to survive. If you cannot bear the grunts, then don't wait. Though not as easy, the best action is to quit before the situation degenerates further. Don't live your life under threat, but don't as well quit a job you love because of a few temporary distractions.

If you choose to stay on working in the same organisation, then you must form a hard skin to such distractions. The hardliners in the company want you to leave your job position for them – just like in government politics. The person who quits because he or she was eyed has lost out, and the others continue. There is someone who wouldn't mind eyeing every other person so he alone will take the mantle. To stay employed, see the big picture, be strong and treat challenges as stepping stones to get to the top. People are watching. Some on your side, some on the other side, and some on the fence. Only you can decide the outcome. Companies like to use tested and tried persons – they are watching your patience. What was a "plan to remove you" may become proof that you are resilient. People get to know you and trust you better. And should some of your old enemies become your friends – for you, the sky is the limit.

3.8 Plans to Get Rid of You?

It would be dishonest to ignore the emotional trauma people undergo in the industry today. The world is changing – the internet and media are linking people with more freedom. Even the pay is no longer as attractive in some oil firms. With inflation catching up and overtaking annual salary increase, it is more difficult to fend for families and dependants with developed appetite. Yet the demands in the office are getting stricter in some oil companies, as more experienced people are seeking the position you now occupy. People face their daily job with mixed feelings. There is a feeling of burden, to the extent that a few personnel have committed suicide because they were not given the attention they needed.

As we go through this period of tension in the oil industry, the need to develop the ability to withstand human-to-human psychological and interpersonal emotional injuries cannot be over-emphasised. When the going gets tough, like animals, some individuals start to prey on others covetously. Most oil companies are large, especially the multinationals, and one person cannot outrightly sack you from office. The plan to remove you will mean setting up traps and situations that put you on edge, until you fall off or any slight wind will send you rolling away.

To shout out loud that you are haunted in the office makes you the first object of suspicion. The workplace has been there for a long time, paying salaries and providing social amenities for staff comfort. The person who complains seems to be the problem. What many don't know is that big companies may lack the basic things in life that we appreciate so much, which the smaller less paying companies allow in abundance. Big companies hit where it hurts, like separating you and your loved ones, disconnecting you from your children, denying your identity, so often, so suddenly. As everyone puts up that winning 'happy' face needed to succeed in that particular setting, there are many who are true to themselves and are feeling deeply hurt and helpless.

I dedicate this passage to those individuals who either due to their own errors or those of others, know that things are not as golden as they seem. It is not easy for them to resign, as you would expect of someone who is not happy in smaller less engulfing companies. They are in the big company which was once their choice, well tied in, alternately weeping and smiling. It is like when an immigrant is told to love a country or leave it; it is not easy because there are citizens who don't like their country and wouldn't quit. I devote this section to you if you feel that no one else knows, listens, or understands your plight.

Back to the staff member who is no longer welcome in his company. Maybe you refused to align to a company cause you do not believe in. Maybe you have become so close to your homeland authorities who are checking the excesses of the oil company. Maybe you are in the contract and procurement department and you have begun to question some kickbacks. You ask too many questions about budgets and invoices. Or you are just not good for the system and your employer is frustrated.

As mentioned before, most of the time, the oil company will not throw out the worrisome employee immediately. There is a fair play culture and individuals are not single-handedly sacked. There is a staff union that may intervene. The oil companies are cautious, powerful but careful. They have an image and their strongest weapon is to be seen as having principles and to be organised. There may be an unwritten plan to remove you.

Often, people think it is the continuous payment of their salary that would be the challenge when the company is in a wait-and-see mode on your case, but your single salary might not make much difference to the company. They may keep paying you even when you are no longer on the credible personnel record, the list not written down on paper. Things will begin to go differently. There are ways to make you put up with their tactical manoeuvres, until it is too late to change anything in the game plan set forth to relieve you. Here's how it might go:

1. You will start receiving contradictory career plans – for example, one too awesome and the other too gloomy.

2. Your hopes are raised about a project, a mission or a posting, then you receive sudden disappointment.

3. You get sudden promotion and demotion. Some positions or moves are not real – just verbal instructions making you act temporarily, like pack up all your belongings (to discover there were no plans).

4. You are cut off from collaboration with others – there is an attempt to isolate you.

5. Someone at the top says you are so important for a given task; he makes you work under him, then gives a poor appraisal and proposes the next plan for you as a correction of your poor performance.

6. You are 'left to rot' in a department where you are engaged in a manual task not adapted to your profile or a job that should have been of short passage, where you cannot possibly avoid getting frustrated after a long while.

7. Colleagues are pitched against you when you are directly assigned misplaced roles now and then. You are often in a position to have to explain to others what is happening.

8. You eventually become 'unassigned', meaning that you no longer belong to any department. Your salary and benefits are no longer justified by a departmental budget. You no longer report to a specific hierarchy, as such no one clearly protects or defends your case. It is as if they company is struggling to keep you going.

There is a common saying that if you make a blunder, "they will send you to Onne to count pipes". Onne is a port in southern Nigeria near Port Harcourt, from where many upstream supplies are shipped to offshore installations by vessels. While it is normal to work in Onne, it is not for a reservoir engineer or geophysicist to go there only to count pipes for years. Such transfer or reassignment is hard to bear by an employee surrounded by inquisitive peers, who serve as both pressure and support groups in the workforce. They say you have been placed in the 'cooler' or 'cupboard'. At this point some personnel become moody, frustrated and ask to quit with a negotiated package.

If you are not good at your job and your character is in question, you may have real problems surmounting the challenges when all eyes are on you. However, if you are confident that this is a set-up, there are steps you can take to remain relevant:

a. The first thing is to look at things objectively, being honest to yourself. Have you stepped on toes? Are you willing to proceed or lay it all down here?

b. Depending on the answers to the above, you may have to humble yourself, meet people, reconstruct bridges, mature and use your shortcomings as strengths. You will have to look at your long-term plans: do you want to be a top executive in your current company? Is this your dream? If yes, remain steadfast and weather the storm.

c. If, on the other hand, you think things are working against you due to certain principles you imbibe, then you may make a choice to quit. Else you may be run down until you lose all confidence and are relegated to an uninspiring point of no return.

I will teach you how to remain humble and still make a living in this difficult time in the industry. One reason employees find themselves in challenging situations in their workplace is pride. The school you went to, your academic level, whose son you are, and all such acclaims may have to remain under your hat if you want to build a career in the petroleum industry. There are ways to manage your success until it is mature enough for the public. When you become a top executive, those successes would be printed as part of your biography in the company bulletin for everyone to see. For now, learn to gradually move one step at a time.

It is not easy to stand out and not be noticed. Every other person wants the good you get. One way to manage your early signs of success is to make others take the glory. Praise people for successes you recorded in the team. Don't worry, anyone who really cares gets to know it was you. Have you not heard of office politics? It has many ramifications and does not have clear rules, just attitudes. Know it exists and be wise not to expect everyone to be straight with you. While I don't advise you to play office politics, I do not expect you to win the fight against them either. It is a hurdle you must cross if you want to build a career in that company. Don't fight things you cannot change, change the things you can fight. Choose what is important to you, don't give up your good principles, and learn to manage other people's advances.

Your highest pain will be that from your conscience, especially if you still fail after compromising your standards. Learn to give up something in return for your peace. Those who claim your success cannot defend it; you gain accolades when you defend success people did not know you owned. Perhaps this attribution of your progress to others is one of the secrets of success in multinational oil companies, and elsewhere.

3.9 Who are They Dealing with?

It is no longer possible to pretend that everyone is happy in this dispensation of the oil and gas industry. Keeping silence could be deafening, in the mist of changing communication, a new culture of freedom and online connectivity that seem to burst apart the world around us. Oil workers are beginning to think for themselves, to cry out loud and a few connect with society, breaking the closed-in circles of the industry cliché with its ups and down, both physically and emotionally.

Some line managers, successful as they seem, are not happy with the company, not happy with top management, not happy with everything. They cascade their stress down to their team members who assume that everything is rosy on the managers' side. Some managers are only working to satisfy an obligation in turn cascaded down to them from a higher hierarchy. The troubles are not about the staff member that suffers it, but sometimes a higher debate stemming from the top company stakes. As they say, when two elephants fight, the grass suffers.

How quickly people step on other's toes is a proof that it is easier to destroy than to build. Even top executives do this to the detriment of their company. If you care for your company, you will hold it up and protect people's jobs. Not all oil companies are shady; in fact, many are doing their best for society. Why hurt the useful personnel that built the company? I encourage anyone attempting to take up a career in the industry to focus on the issues that can be improved. Some oil companies are doing their best to provide energy and serve the common good. Few persons thwart the system and use their positions to hurt the weak and foster divisiveness. It is not you they are dealing with, but indirectly the company's interest.

Do we go against every oil company and the petroleum industry at large? Not at all. We should set it aright while working to offer alternative energy solutions. Everyone in politics knows that it is not a solution to destroy the system in place, without an alternative to replace it. The same nature of man will hijack the new situation and install a worse system. We may challenge the order of the day in the oil and gas industry and criticise it, but those who do not offer alternatives are not helping matters.

When some regressive mind pulls your career back and hurts a potentially useful employee, they make the company lose some of their best personnel, first at heart and sometimes for real when they quit the job. If you are good and somebody is suppressing your progress, the company is losing your full contribution. Simple altercation between staff and management has led to major catastrophes in the industry, with millions of dollars loss to the company. No one likes working in a team that is unhappy; with health and safety as top concerns, regressive minds sometimes pose a major risk to the oil industry.

3.10 Progressive and Regressive Minds

The *progressives* move, the *regressives* stall. The progressives are pushing you ahead, the regressives want you to wait, stop, or even regress. Moving is not always the best thing to do, if what is in front of you is a pit. At times a little shake from the opposition is good. Both regressive and progressive characters play an important role in your career. What will push you out of your comfort zone is discomfort. Do not take career challenges personally, or have grudges against one person who tends to frustrate you. It is a bigger game. Sometimes, your immediate boss is just acting on a clue he got from the top.

There are persons who serve as a ladder to your success and others who serve as a trough. You will begin to discover this later in your career, if not earlier. The troughs are probably dissatisfied with their own progress and eventually transfer their pains to a few others. You only happened to be on their bad list. These are regressives to the whole system; they see things differently, often picturing fellow colleagues as liabilities.

One new staff member was once told by a senior that everyone was pitying him and only just managing to retain inexperienced new staff like him in the company.

The new staff member was downcast, surprised to hear he was not appreciated; he lived with this for many years. It was only later he understood that situations are interpreted differently by the progressives and the regressives. The senior staff member was hearing and interpreting woes actually in his mind. The junior staff member discovered that when there are problems in the company, everyone carries his as if all is well. He eventually learnt not to take it as a personal tragedy.

In another situation, a new recruit complained to an older colleague about some wrong treatment from his manager. The old colleague told him to write a petition to the top management. The new employee did so and was fired for writing a petition in his first few months of work. The old colleague narrated this event with a sordid grin, almost happy the new staff member was sacked. It turns out that most of the persons you will contend with in the workplace are your colleagues, those you eat and drink with and smile at every day. Your day-to-day problems will come from peers, not management. They are not necessarily deemed wicked, just inadvertently using you to test out their own ups and downs and transfer their aggression if you are naive.

When you get to any of your work locations, there are those who make comments that lift your soul. They think of the common good of everyone. Surprisingly, they are not always at the helms of affairs; you wonder why. The few progressive minds at the helm receive many critics and their good intentions are interpreted as weaknesses. These are the progressives; they will help you cope in tough times and give you hope to continue. Learn to work with both the progressive and the regressive. They are both important. If a regressive mind gives you a good mark, everyone believes it is good. If a progressive gives you a good mark, people want you to prove it.

Your scoring good and bad is a good thing for the company system. It keeps people striving and never getting it right because a regressive will often give you a bad mark. Progressives are wishing things to go well for everyone, they tend to move you ahead, but in the long run everyone cannot just move all at once. The progressives make you the best you can be; the regressive challenge you enough to see ahead within or outside the company. It might not be those actions you like that will challenge you enough to learn the system. Understanding the progressive and regressive methods make you look at the big picture and smile.

Progressive methods

- *Move you quickly.* They help you leave the base – starting at the bottom can be challenging (you are the stepping stone for many). Appreciate this quick move, even if you later need to go through the missing links and conquer any step you missed.
- *Encourage you.* This is good for the morale, but sometimes you may keep waiting for what may never come. They may give you hope where there is none.
- *Give you a sense of belonging.* This is good as well, yet overly thinking you are the centre of the world may be harmful. Some personnel begin to get so comfortable and think the company is their own personal business. Such can hinder you from reaching your other goals. You need to learn to detach at some point, depending on your career objectives.
- *Trust you.* They allow you space to use your initiatives. Trust builds capacity and confidence, since without confidence in yourself, you will fail when asked to do even the most basic tasks.
- *Tell you, you can.* They advise on your weak points and emphasise your strengths. You need to hear such soft advice; your weak points may be more than was mentioned.
- *They cross levels to reach you.* It is important to know that someone cares, especially those at senior

management levels. Progressives leaders are open to all staff and past team members.

- *They see you as great.* They see the good side of your future and think of you in the long term.

Regressive methods

- *They kill with words.* They leave you with a few words that infect the heart. Since people are trained to be respectful at this level, it is the little words that kill slowly. Never lose your self confidence.
- *They are hard to please.* They change your duties or position if it seems you are doing well and others are appreciating you. This may well expose you to many challenges and serve a positive end: you get to know the jobs and the people, you become used to working with many different characters.
- *They emphasise your weaknesses.* You may be engaged with complex tasks to highlight your weaknesses, or they leave you with an unimportant text editing task for a long period, tending to run down your area of competence overtime. They let your peers know.
- *They get you destabilised.* Be wary of changing boundaries, changing rules, moving standards. If someone changes the order of your presentations at the last minutes, maintain your message and not get destabilised by attempting to follow his thoughts.
- *Appraise you poorly.* It is not just the appraisal result that matter, but the method. They try to alienate you from other persons who will assess you so as to have the final say.
- *Castigate openly.* Good bosses don't; they deal with issues, not with the person. Regressive minds talk to others negatively about you and try to follow it up, leaving reports and plans to stagnate your career when handing over to a new manager.
- *They have an alibi.* They work with the newest persons in the department, having lost the confi-

dence of others. They always have a favourite and a scapegoat.

- *They see you little.* They think of you in the short term only.

Being under regressive control for long can be daunting. You will need a new manager with a free mind to break out of poor appraisal; most managers just refer to the past to give you a new rating. The overall comment made by the previous manager for your long-term plan will be used by the new manager to rate you. The good news is that the way the oil companies are organised does not allow room for long-term prejudices by middle and low-level managers. The oil companies' operational structures keep changing your position and location; they only leave you to carry over your indictment if your problem is truly affecting the company. They offer you an outright wrong placement only if they want to punish you.

Progressive and regressive minds are like two teams on a football field; while they compete, the company or owner of the stadium they hired is in a win-in-any-case situation; he earns irrespective of who wins individual matches. It is like two sides of the same coin. Like the Republican and the Democrats in the American government. Only you can drive a change, for yourself and for the larger society.

3.11 The Later Part of Your Career

You will be caught up in the company system and industry antics; you'll forget you only wanted to make a living when you got that job. People outside have lost connection with you. You are now of age, and you find that money does not solve the inner problem of fulfilment and guilt. Some personnel on retiring resort to spirituality, for others it is too late to change their ways. Managing the later part of your career is actually managing the later part of your life, planning for old age and achieving a life purpose.

Should you be in the oil and gas industry for good, then you will have to work out ways to be happy, share the blame, and at least survive. You can join the well wishers, the progressives, and contribute your quota. The efforts of many honest people are why we still have some level of sanity in the industry. You need to understand how success works and when money enslaves, then master the comfort items. To start with, if you want to become independent or happily retire some day:

- Don't buy comfort at all cost, to the detriment of your financial plans for the future.

- Invest more in fixed physical assets (like rental property, where you have the same building under your control even if rent amount changes) rather than floating assets (like stocks where you have no physical property and do not control the price).

- Don't count on financial experts to do the investment for you, but learn and take out time to participate personally.

- Spend time with family and loved ones and with yourself, who will matter most at all times.

- Don't discard external relations and isolate yourself from friends; don't insulate yourself from problems in the society. Begin to solve some of them.

- Believe there is a purpose for your life and destiny – then pursue it.

Take the many honest pieces of advice about being true to yourself and adopt the strategy of investing for freedom which you find later in this book. If climbing the corporate ladder of the industry is your aim, make sure you are happy and fulfilled inwardly. Understand the functioning of the corporate organisation in which you belong, early enough. The time to start is before the later part of your career: start today.

Chapter Four

HUMAN RESOURCES AND INTERNAL FUNCTIONS

4.1 HR and Administration

The Human Resources (HR) is the government within the company. They oversee the employment and development of staff career. They liaise with administration to follow up work progress, rank personnel and decide on promotion, and sometimes demotion. The size of the HR department and the scope of administration is closely tied to the available funds, managed in collaboration with the finance subsection. The finance department is checked during auditing performed by qualified accounting personnel.

Internal auditing is separate from external auditing, which is performed by representatives of the national authorities. Internal auditing checks excesses in expenditures, in the interests of the company, and ensures that the finance department is accountable for and ready for any outside checks by external auditors. They put the house in order. Many staff do not realise that they form part of the expenses and are liabilities until they are able to work out their pay.

Another HR subsection, charged with methods and procedures, checks that there is compliance in the technical and administrative roles, while also ensuring that the company meets the expectations of regulatory bodies. In oil and gas companies, the career advisor's or career manager's roles are between technical and administrative duties; sometimes the advisors are former technical personnel who have evolved to now become professional guides for staff. That is why they understand the challenges that employees in the various departments face; they have an overall picture across disciplines. Their duties are:

- Facilitate career growth and exchanges among staff and managers.
- Encourage and resolve issues between technical team and the administration.
- Identify the needs, organise and catalogue the training of staff.
- Foresee recruitment needs and ensure quick and suitable replacement of personnel when necessary.
- Develop career plans for personnel in line with company objectives.

One useful role of HR, to a young staff member who has an issue, is that he or she can run to a career manager and the problem will be looked at from outside of the scope of day to day work. Though they speak to your hierarchy to find out about the job, a good HR team would tend to address your challenges from the social side, overall work and life balance, including how your health and family affect you on the job, and vice versa.

4.2 Main Duties of the HR

The HR introduces new recruits to the company, follows them up and keeps their database among that of older personnel. Their training involves a study of employee tendencies, which is why they look out for behavioural excesses in the office environment. The HR keeps remuneration and benefits in check, just enough to keep the workforce going. Everyone is kept happy or made to feel a sense of fairness. Although not always the case, there are guidelines for career growth, skill development training, managerial training, promotions, long-term work plans, replacement plans, gratuity, pension, health care, security and welfare.

These provisions may come with escorts, good dinners and lunches, peer competition, comforts, loans, large pay cheques, cooperatives and all the liabilities that come with the good things of life. In listing the above, notice how we move from the essentials such as training to peer pitching among each other, resulting from HR ranking different welfare packages for different grades of staff. As colleagues work towards the conquest of the HR offers, they are entangled, soothed, calmed and tamed by oil money, as if completely slimed in cold waters. The challenges divert the personnel's attention and render them docile. "Why does pouring oil on the sea make it clear and calm? Is it for that the winds, slipping the smooth oil, have no force, nor cause any waves?" – *Symposiacs, book viii. Question ix.*

Filling in of forms, ticking time, and staff expectations come with the fear of disappointment. The nagging fear of the opposite of what is desired, such as insecurity, old age, poverty and death cause more harm when you are not in control of your life. This same fear makes you keep losing control, and keeps you in check to remain within the organisation. There is not much to be offered to you by society. Oil workers in developing countries seem to have no alternative panacea. Society is looking up to you, so you may not have the option of looking up to it.

The main responsibility of the HR in major oil companies, in line with their official duties, is to retain the staff and to find the just milieu between remuneration and work results. To add benefits and packages incentives to the workforce, which come with documentation and tracking, performance and controls, such that the gains must be strived for in a setting where some will obtain more of them than others. The ever increasing efforts to equalise keep the working dynamics where people are well bought over for life.

4.3 Recruitment and Manpower

The recruitment and manpower subsections of the HR determine the strength of the workforce with respect to a workable payroll. The payroll is the number of personnel on the employment register and the amount they are paid every month as salary. The HR attempts to find a balance between meeting company manpower needs as well as ensuring a sizeable payroll in a financially manageable proportion. There is a good tradition of positive communication to the public that goes with the recruitment of best candidates in the industry. The idea is that the mindset at recruitment determines personnel commitment for a long time. The success of a company is determined by the quality of staffing. Your mindset at the place of work is key.

The use of external consultants and recruitment agencies help companies to be more objective and ensure there is no bias due to conflicting interests, tribal or ethnic affiliations, especially in the African subsidiaries of the multinationals. The select few who get through the agencies are then revisited with interviews and in-house technical assessments before final employment. Within the company, number of employees, their roles and evolution remain dynamic. Staff strength reduction is one of the first areas of adjustment when there is an economic reorganisation in the industry.

Disgruntled personnel are a risk, first to fellow staff and then to the company. The HR does everything to filter out persons with such potential of rediscovery during recruitment. They keep refining the relationship between staff and the company with awe, doing that wonderful work of keeping the smartest people in society working arduously with utmost docility, as if they were born to search for oil. It is as if the HR is constantly asking, did we get it right? Any square peg in a round hole? Everything must be maintained to keep the wheels turning.

4.4 Checks and Balances

Human resources are not happy with managers whose team is not performing well at work. They provide training for managers that moderate their ego and improve leadership skills. The aim is to make managers think as the binding force, responsible enough to bear the burden of continuity for the good of everyone. Oil company managers are well remunerated and compensated, to remain loyal to the work system while driving performance in their team.

The HR personnel themselves receive training in company organisation and behaviour of people. Where the 'progressive' and 'regressive' attitude find its way into the core HR team, the long-running company system, and interventions by the top executive and sometimes CEO, resolve the issue. You therefore find yourself in a setting where everyone is busy doing different things, but all work somehow in the direction of the company's corporate existence.

4.5 Career Follow-up

The human resources rate staff performance over a long time period. They get feedback from managers without needing to understand the nitty-gritty of the tasks. In a good HR system, an error made by a staff member is cleaned off the file after three to four years; only the last few years are used for short-term rating. They analyse why a staff member behaved differently in different situations. The HR also evaluates bosses, why one boss got different performances from staff compared to the performance of staff while under another boss. Indirectly, both the staff and HR are evaluating the boss as he or she evaluates the staff. A boss getting good results is also a boss performing well; a boss who always has issues with his staff is a boss with issues.

A company that does not invest in career management cannot survive the new trend of the industry and the world at large. The new generation of personnel is smart, internet savvy and knows what you know. They like the good life, freedom and enjoyment, but hate closed-up office spaces. You will have to be neither too soft nor too hard on them. A modern career plan attempts to integrate the personnel's career dreams. The company that doesn't will have to compensate a lot with financial rewards, which alone is not enough for employees of this age. To survive the test of time, you must keep reassuring the personnel that there is no conspiracy behind the blind, that there is complete transparency. The unions could be ready to throw all away, and the company has to be strong enough to either break such a union or survive them.

4.6 Propel Your Career

A manager pushes staff to the left and to the right. In the end, if you did not give up, you ended up moving ahead. You will work with other people, time will pass, people will move. Challenges end up highlighting good quality, patience and resilience, which will propel you into the limelight. "You cannot hide a golden fish," was a popular quote by my mentor. My mentor told me that it is those who don't like you that will bring out what you need to propel your career. Those who like you will attempt to make things easy for you, pushing you too quick that you could end up making costly mistakes.

Those who don't like you make you go through all the processes and beyond, strengthening you inadvertently. The company does not get rid of the wicked; they place them in strategic positions. They are sent to destabilise a team that is getting too comfortable; they are used to penalise defaulters. They are tools, unbeknown to themselves. Often, they work too hard to gain attention, but only receiving a little. They are used as a final test to highlight a winner.

It might not be what you think, as there could be a bigger plan lurking behind the challenges you face; when it turns out that one person is saying you are too good and the other person is saying you are too poor. When the HR transfers you from here to there, changes your programmes now and then; even those in the travel department can predict that a new executive is born and awaiting time to unfold. Many staff members do not get it at this time, they become bitter and lose the opportunity. What will propel your career at such a time? Character, loyalty and availability.

Sometimes the game is not worth playing, and the loyalty aspect gets in the way. You need to evolve; these three parameters: character, loyalty and availability, will affect your evolution. It is not those who come in and seem to know it all at the beginning of their employment that will ride to the top. It is those who may be timid at the onset, but who evolve to be the average persons in the office, and then supersede. You need to be present and known by the top management during this evolution. When there is a meeting, a decision, be there. The story always ends with the person who is absent being blamed. He who is absent will have to defend himself later. There will be times when things will make you want to run, to fight, to quarrel. The final characters in industry growth will just absorb it, pretend it never happened. That is the price to pay.

At the early stage of your career, whether you have the company's interests at heart is not very important, or obvious. Young staff usually like their company. At this stage, a good character is already important. At the second stage, as you progress, a second factor comes into play, which is how loyal you are to the company's agenda. People who have the company's agenda at heart are very available, even to a fault. They work extra hours, weekends, late meetings, in-corporate parties; they spend all their time being visible to show they have nothing else to look up to. That corporate ladder is presented to you to climb when it is obvious that you are dependent on it and that there is no other one. It's like marriage. You are married to the company. This is your farmland, where you earn your daily bread.

Many employees do not want to give in that much, yet they expect more in terms of career growth. They want to be placed at the top of the ladder, because they are smart, intelligent and hardworking. However, it is more than that; the consequences of climbing the ladder is also the willingness to bear the loss and to fall with it should it collapse. You don't lean on another platform and you want the ladder placed before you. It is not the problems you meet in the industry that are the issue, but your reaction to them. While you narrate, people are not listening for who is right or wrong, but what most represent their interest. The company prefers to lift personnel who they have ramshackled, sifted and tested. They promote you when they are assured that you would stand the test of time and have the long suffering it takes to be a top executive in one of the most challenged industries of our time.

4.7 Organisation of Work in the Industry

Your manager is not the only one in charge of your remuneration or even promotion. Some of your requests will go directly to the HR or Finance, or to a technical representative for your work division encompassing many departments and managers. In most cases, however, your manager will have to append his signature, which should not be unreasonably denied. Every entitlement applies to other colleagues and you are treated in a herd of different groups or grades. The organisational framework in most oil companies ensures that there is no one single boss, expert, option or decider for each personnel or job.

The supervisor reports to the manager, who reports to the general manager, who reports to someone else in the upper hierarchy, but there are rooms for horizontal communication, open discussions, meetings, joint reports and cross-functionality among the levels. In some oil companies, any staff can walk up to the directors or fix an appointment to have a word with the Managing Director or CEO. That doesn't mean they do what you want. If you are an expert, you find out that the expert technical result is not the only consideration. There are political, environmental and cost-versus-partnership decisions. There are always different options, the most likely, the unlikely, the very likely, and sometimes one being the opposite to the other. You are helpless in terms of control, as in a plane! The system is stronger than any one of its collaborators. The daily work routine looks normal and no single incident of turbulence can crash the plane.

4.8 Rotation and Shift Duties

On rigs, where the supervisory roles are more direct, there is a check on the tendencies of supervisors to maltreat the persons in their charge. Shift duties give each junior staff member the chance to do their work independently and under different supervisors. Emphasis is laid on the need to work together in harmony to survive the harsh rig environment, especially offshore. No one wants to annoy someone else so much that a person is pushed into the open sea! No wonder, the psychological tests you undergo during the employment process.

The rotation and shift approach creates a mindset beyond actual work at the site. It is a tendency to ensure that the only stable thing is the job. Every other thing revolves around the objective. By varying the personnel, the system remains constant.

4.9 Transfer and Regular Moves

Transfer could be from one installation site or office location to another, or a cross-posting from an affiliate office to the headquarters or to other subsidiaries. Relocation may also be from headquarters back to the affiliate office. One wonders if these movements are done for career experience or again for checks and balances, so that the staffing keeps changing while the work remains stable. In addition to what has already been mentioned, such as ensuring that no one single manager is controlling promotion, and that no one single staff is victimised for long, movement of personnel also ensures that:

- Build up of stress is released pending the build up of a new one.
- Each change comes with its new beginning, encompassing the staff.

For a young staff member, this may seem a wonderful opportunity. You are exposed to new challenges and learn to work with different people at both local and international levels. You gain experience in different fields and regions of the world. When you start to build a family, you begin to see that stability in your life plan might be affected. Children's education abroad may be subsidised to cushion the effect of disorder in their upbringing and to encourage the staff. The other less visible consequences of staff movement are:

- Movement from home and of properties on a regular basis.
- External investment of the staff fail when displaced
- Staff becomes over-dependent due to overly helping hands in the HR during cross-posting.
- Home built by a staff member in one place is left fallow during movement.

In some oil companies, a senior staff member must move every three years. In a foreign affiliate, some staff members need help with everything: children's schooling, taxes, and the only investment they make is on the company stocks or cooperative. In the end, your life is like a package in the hands of a courier delivery service – ready to be sealed and posted to the end of the world. The company feels justified by the huge amount they pay staff during cross-posting and as such expect full commitment and service. Now that there are pay cuts, such commitment is in jeopardy and there might be a need to rethink the relationship.

Regular movement of staff has other upsides and downsides for both the company and the staff. With the exposure, a staff may resign and join other companies during international assignment. There are huge costs and administrative challenges, including those from the government and immigration authorities. Health issues and eventual death during posting in a foreign land is a tough challenge for everyone.

Meanwhile, international movement to the headquarters is a time to check a staff member's competency as well as his or her character. You are allowed to be in a relatively free operational mode, where your true character emerges. It is also a test of assimilation to a different culture and tolerance. Will you, during this period, understand how things work? If you are cool after this time, you are cool forever. When you return from a long-term mission in the headquarters, few things happen. It is either you are:

1. *Demoted* – not necessary in your job position or visible grade, just a change of outlook on you. You may become a suspect for having shown doubt.

2. *Promoted* – it might not be immediate on your position, but you are now dining with the great. Managers refer to you when talking to other colleagues, gradually building their minds to accept you as part of the management.

3. *Move laterally* – 'lateral movement' is a particular case where you do not fit into any position and you are also not a threat. Might need more time to observe you while you are given a similar role and placement to the one you left.

No wonder the industry is proud to present many locals, who have had international exposure, who now head the states of affairs locally in their country. Truth be told, most of those people were either so submerged in their work and challenges, or were too tied to their daily bread which their disturbed families will forever need, even the more, to really play a constructive role. There are but a few who will make a difference – and experience has shown that a long-term plan involving scholarships, actual integration into foreign society on a neutral ground, before working in the foreign headquarters will do more good to the nation than just short-term work cross-posting.

You need to go to these foreign countries on other levels outside the corporate facelift of the multinational oil companies, to understand the psyche of the common man in the Western world, to be able to reflect on the context of headquarters-to-subsidiary relationships when back home in Nigeria. The normal cross-posting is so official and stage-managed that it does not expose you to the true scenario. On the contrary. When a personnel develops an independent mind while at the headquarters, then returns to Nigeria, his mentality shocks the rest of the in-country expatriates who are not used to Nigerians thinking and acting that freely. They tend to see that person as a misfit. Here then lies the greatest challenge for your multinational employers; why should they train you if they won't retain you – body and soul?

4.10 Money, Retirement and Old Age

Consider the later part of your career, before it starts. When you see aged staff in the oil company offices or on rig sites, you wonder, is their retirement not due or are they still there because they so love the job? Stressful working conditions can make one look older than they really are. Working until you are 60 or more sounds strange when you consider that these people get so much money each month in the form of salary and allowances. Where has all this money gone? Couldn't they have quit and now be avoiding the daily traffic and benthic office 6 am to 6 pm daily routine? The job could take so much from your life and sometimes life ends as soon as you leave the system in the form of retirement.

People think you are rich because you work in an oil company, but surprisingly, the money lasts only during the time it is being earned. Except for a few who had a plan early enough, it turns out that money alone does not make you rich. They have money but no time, which makes them solve problems by "sending money". Money cannot, but is often used to, replace time. If you have so much money with less time, the money will be half as useful. Time is what many lack in the oil industry. Contrary to in-house work achievements, oil workers do the same out-of-work projects over and over and hardly ever complete them. Even with rotation, holidays and weekends, the nature of the work is mentally demanding; you seem to create a parallel work when away from work, requiring high concentration and follow up of a running workflow all the time.

Those who are on time-off from the rig restart all they have built while absent, at zero. Most of those things were never well done. Those who work in the nicely set towers in town, badge in each morning, walk into an edifice and face the computer all day, with limited communication with the outside world. They develop a solo life, with any social life being colleagues and meetings. Few will meet up at weekends, or meet other people and communicate with their loved ones. Even fewer will run an investment to keep them going if they stop work. Oil personnel focus on investments such as shares managed for them by someone else. That someone makes the profits first. There is a challenge of integrating into society when oil workers quit their job. Some have family challenges that worsen – leading to divorce, not just with their wives but the entire household.

Questions have been raised on why major oil company staff die just before or soon after retirement. Naturally, younger staff are concerned. The health status of those field personnel, changes in their body system resulting from exposure to gas and chemicals while offshore, and work stress are suspected as some of the causes. Few worry about the lack of purpose and possible late-age crises they find themselves in once out of the very work system that has bought their soul all lifelong. A sense of emptiness, lack of planning and a sudden relief of life pattern takes a lead in the cause.

At this stage, you will begin to understand that the personnel in the industry deserve what they are paid. That is why they have health care integrated into their career and work life. In some countries, health care and insurance is 100% for staff. In the oil company, there is the good salary to compensate for any of the other things you give up, and most especially a good pension plan so that you do not have to worry about your retirement and old age. That these benefits are changing is why we have to discuss them now, to give current personnel a way out.

Today, the contributory pension arrangement and drawdown in national social security creates tension. Some people still hope to earn their pension; many others are having doubts. The idea now is to attempt to work towards breaking free at some point. However, it takes some work to set your future aright: *plan from day one.* Your day one could be today:

- *Add personal investing to your contributory pension.* Would you rather invest in real estate, paper assets or a business? All demand extreme sacrifices at night, lunch breaks, weekends and holidays.

- *1st year is ultimate.* Your first or early years in the industry are crucial; if you fail in the first year, then carryovers impede on you excelling in the subsequent quests. Don't live too big too quickly.

- *Avoid having a distanced wealth.* Visit and be close to your assets. Else you have it but can't control it. While you spread your efforts, choose a town or

state as a base to maximise investment and somewhere to settle.

Most of the regular industry personnel you admire
today are in a dilemma. They are in the middle class; they
have money to spend, but are tied down to stressful daily
routine work for life. The real rich people in the industry
are the inside investors – some of whom are non-visible
executives.

Chapter Five

SELECTIVE RECRUITEMENT
AND CAREER PROFILING

5.1 Oblique Relationships

There is a debate as to whether the oil industry major corporations try to maintain a perpetual hold over their subsidiaries in developing countries. According to this argument, the colonial approach of 'divide and rule' leads to chaos in the subsidiaries. The host nations would have to sacrifice a lot of interests in their bilateral relationship with the IOCs. While the local collaborators are tussling for recognition, the oil resources are harnessed and external supports go only to the local party that is willing to cooperate.

It seems there is a system, a well-oiled machine, refined by centuries of experience and colonial culture difficult to counter. In my other book, *Overview of the Petroleum Industry*, I defined the 'system' and the principles behind it, showing how it is strengthened by survival tendencies of individuals, companies, nations and the world at large. The principles behind the system controlling the oil and gas industry are rooted in the most innate nature of man: self preservation, making one party to override the other in an effort to survive. Perhaps it was that same principle that led to colonisation.

In the sharing of limited resources, interest always overrules. People do what they do to keep the oil industry afloat, their source of daily bread, amidst the many challenges. The system ensures that people invest and obtain the proceeds, sometimes to the detriment of others. Many work directly or indirectly to maintain this system of control that maintains the flow of resources and development in a given direction. This is the system you have to serve if your expectations of career success in the industry is to gain favour and financial rewards.

5.2 Questionable Outlook

Over the years, it seems the way the African nations are made, or have come to be or behave, makes the 'visitors' enrich themselves while Africans are impoverished. When acting on international matters, we tend to take the role of visitors both at home and abroad. Persons in developing nations see business associates as people arriving to solve the nation's problems, while these business associates think repeatedly that they are going to a farmland. They come to work the land and the end goal is to harvest. It's not an issue of right or wrong, it's an issue of survival. Your work in the oil and gas companies is somewhat that of a farmhand.

We do not always remember that people develop a farm to harvest it, so we focus on development. Our investors from across the Atlantic or Mediterranean always want a contract to be signed. Even slaves were paid for in those days; it could be a simple piece of shell, but they were paid for and as such 'legal' in that sense of it, which still legalises unfair advantages to the domination of oil majors in developing host nations. In the times of slavery, it was the duty of the seller to be responsible, to choose for himself and not sell his brothers and sisters. The seller then justified his actions, such as the possibility of finding greener pastures in the land of the masters. Are we making justification for similar reasons today?

With the quest for a larger share of food, and other wealth and riches from nature's resources, comes some of man's worst inhumanity to man. To make ends meet in this high stake role, some core human desires, ranging from family ties to peace and true freedom, are gradually being ripped away and people become increasingly depraved. It is a trap in which one sells his soul for a few pieces of silver. It creeps upon one slowly, then it is too late to retreat. You must keep this in mind as you pursue your career in this domain.

Whilst in Europe, I noticed that friends were wary when you wanted to buy them a coffee. "Why are you doing it?" they would ask, as if there must be a reason for every gift. Paying for someone gives you an edge. It is like giving a dog a bone to tame it. So they say, "No thank you." If by chance they borrow a cup from you, they soon pay you back and remind you that they do not owe you anything. I learnt from them. Back home, my people assume that the other person is here to help, but I found that people I met abroad behaved as if the other person was there merely to complicate issues. We often see the image peddling side of our foreign friends, and mistake it for care. It could be, or not. In France, they say, *"se méfier..."* a kind of *"be careful"*, don't be quick to trust. It is not about you wanting to be kind; it is about understanding how the world really works.

In Nigeria, I find that some of the expatriates called upon to resolve issues on fairness were the very persons that created the injustice in the first place, unknown to our local personnel in human resources. They assume that since the model of the expatriates works in their country, then the expatriates are here to apply the same model. Unfortunately, even those expatriates who have good intensions are gradually tempted and eventually succumb to the naivety of some of the local elites in the company.

I find a major difference in attitudes between the African side and the Western world, as shown below. Although I do not advocate the adoption of the contrary attitudes, I hope we understand the consequences.

Our Attitude	**Their Attitude**
If something belongs to everybody, it belongs to nobody.	If something belongs to nobody, we are here to discover and possess it.

A visitor is an angel, we give up our comfort to please them.

A visitor is a stranger; beware until we have known them well.

The new person may have the solution to our needs.

The new person may take away the current resources.

We accept the visitors, who should have good intentions. If things don't go as we expect, we fight; we show them what we are made of.

We are not friends, just business interests. If you have the resources, we hold our technology. We remain cordial in a win-win situation.

The successful and capable will help the poor, capability comes by luck.

The rich indirectly takes from the poor, who do not know the value of their own resources.

Follow whoever gives you food, the provider has the right model.

Why should you feed us? There are no free gifts; we prefer to be independent.

Please be kind to us, we trust you will not cheat and hope to get your favour by so doing.

If you accept our bargain, then the business is legal and we do not owe you more than that agreed.

Everything is right or wrong; high calibre persons would not do wrong. God will judge the wrongdoers; whoever does wrong gets punished.

Right and wrong depends on the context. What people think matters. Strategy, not emotion, counts. He who plans better and acts accordingly deserves to be rewarded.

One-off financial reward, status upgrade and material gifts compensate

We are interested in the long-term plan. Contractual resources or cash flow, for giving away what we have.

leading to steady even if minimal income.

Follow the patriarchs, the ancestors; youthfulness is exuberance and has no say in society than to obey.

The future belongs to the youth, give them the chance to innovate and lead us into the future.

I remember the first time I travelled to France. On our arrival, we saw that tourists were either appreciated or ignored. Once settled as residents, we were surprised to see that the appreciation tourists got was not for those who were coming to stay. There were initial misgivings as we struggled to make friends who ignored or discouraged us from continuing to try. In the training organised by a corporate French organisation for us, we were told that the French are like coconuts, that you have to break to find the juicy inside. So we worked hard for acceptance. You won't get to know all you want too quickly. They hesitate before they answer your questions. The act of learning about someone and being learnt about is costly, an art in France.

Then the fluency in speaking French was judged as one's level of intelligence. Though our French friends could speak English, they hardly did so with the foreigners who had to struggle to keep up with them. If you disagreed or argued with especially poor French expressions, then you were tagged as dissenting. The French word for foreigner is *"étranger"*, stemming from *"stranger"*. Foreigners it seems were treated like strangers in France until proven otherwise. Of course, if you were coming from the USA, they thought they already knew what kind of stranger they were dealing with, and you would be welcome more quickly. On the contrary, most Africans welcome you first and then try to get to know you afterwards.

I also find that seniority supersedes a sense of judgement in some African cultures. Elder brothers or sisters could whip and hit their younger ones as often as they liked. The parents condoned it, and often told a child to be more like their elder brother or sister. Elder brothers were always right and words from parents to siblings show that they did not expect the younger ones to be better. Yet most successful people I have met, at home and abroad, have the mindset that their ideal life model was not their brothers and sisters, or even their parents, but someone else.

Seniority in terms of age, instead of merit and competence, has decided who rules in most parts of Africa. Clueless old persons have been maintained in office as presidents because they are revered as kings, irrespective of capacity to drive the nation ahead in a world evolving with some of the smartest young lawyers as presidents elsewhere.

5.3 Docile Proxy

Our schools teach the colonial culture in colonial language, presenting it as superior. We grow to develop a lack of interest in our own culture. We are like proxies, looking over our shoulders to some destinations away from home. People engage in proxy roles when they play the second fiddle. In Nigeria, proxies are the ones that are sent to negotiate with the National Petroleum Investment Management Services (NAPIMS), and carry out negotiating functions. Poor images at the national, corporate and socio-economic levels, is exploited by foreign investors for their own benefit; their careers naturally supersede that of the local elite. A colleague in senior management once told me that we are running with persons on horseback. We apply the "Who knows who" principle, where those who give eye service benefit. The oil lords adopt the proxies and together they wield authority in matters that concern our country.

As you pursue a career in the oil and gas business, especially upstream, you must be aware of the challenges related to where you come from, learning to not let it limit you. The times call for local persons in the host oil nation, most of which take competently normal roles when abroad, to be able to run their own resources back home. It starts in the mind, knowing we are capable, acting accordingly and showing it to the world. Online activism is developing and it seems that the new generations are beginning to ask such questions that make a people to stand out and demand their rights, most importantly, to do the necessary homework to be able to manage those rights. It is lack of capacity that makes active professionals to serve as proxies to the farm lords instead of taking charge. We have our own part to play, striving to be good in the technical aspects of the business at all costs.

5.4 Meritless Principles

All through my experience in Europe, though there might be occasional denigration of merits, it was never with impunity. Back home, we place aloof 'monarchs' and usually retired persons in top positions of leadership. Our leaders act like unopposed gatekeepers of the nation's resources. They are compensated by merchants interested in accessing the gate. Citizens have become comfortable in suffering, creating comedies and patronising stand-up shows to laugh away their sorrows. Our resilience has been in our suffering, not in the painstaking to find solution.

We avoid smart people as leaders to avoid having issues. In a country like Nigeria where the innate fear of division is sublime, suppression of youth awakening by the old rulers through active and passive means holds sway. As the free mind is suppressed in Africa, the creative mind is also tranquilised. That is why though Africa holds about 70% of the world's natural resources, we process less than 2% of it. Fear does not make people to act well willingly, only when they are monitored. This is why corruption remains endemic despite a seeming sledge hammer on it and the anti-corruption propaganda.

While we recognise this lack of efficiency in our local processes, there is also a tendency to exaggerate our shortcomings. There is a tendency to criticise the less advantaged or developing nations and their people, while presenting the advanced foreign business owners as messiahs. Some NGOs, religious organisations, the press and expatriates, try to show the Western world as flawless. When whites come to Africa, their application of merit is no longer the focal point, but seen as "people we can work with". Unlike in the Western countries, where the multinationals are strictly checked, some of them don't apply merit in the affiliate countries; interests, more importantly, that of maintaining a working relationship between them and their host countries or communities by settlements, takes priority.

The most surprising to me when I returned to Nigeria was the absence of that free thinking, open debate, and free lifestyles in the workplace compared to what existed while were abroad. Here, those same colleagues now frown, take respect as too important and lobby at work. One of Matthew Kukah's controversial lectures clearly explains why this happens.

As reported in his column in the *Punch* newspaper, he said:

> If you take the worst Nigerian President and Americans are patient enough to vote for him and he rules America for 20 years, he will not be able to steal up to $50,000 no matter how long he rules because of the kind of system that operates there. But if you take a British Prime Minister or American President and make him the President of Nigeria, the moment he wants to set up an anti-corruption

agency, he will have to bribe the National Assembly.

– October 2, 2015 : Eniola Akinkuotu,
http://www.punchng.com/politics/Buharis-Ministers-
Must-Declare-Assets-Publicly-Kukah

A point I could pick up there is that people may change and cheat you if your system permits them to. So it is no wonder expatriates in Nigeria do not behave like this in their own country. I see some young graduate expatriates being high headed with a special feeling of specialty, wielding commands and demanding respect. Most of my local colleagues could not tell the difference between the experienced expatriate who really had something to offer and the young expatriate who was also a learner. There was an amalgam from the local onlookers, that everyone coming from across the shore had the solutions. While in France, everybody knew that there was the *"gentil"* and the *"pas gentil"*, that is, *the kind* and the *unkind.* An important thing you needed to know before dealing with each person. You are trusted according to if you are *gentil* or *pas gentil.*

On the other hand, returning back home to work is a challenge for young Nigerian recruits from popular universities abroad, who not only travelled for their studies but grew and studied outside the country. The principles they imbibed, of freedom and free attitude to work, isolate them. Their sponsors want them to know that they belong to this side of the fence. When you behave in a smart kind of way, walking fast over a very short distance, speaking with tension in your voice, to show seriousness, your local colleagues scorn you and imagine who the hell do you think you now are?

It took me a lot to adapt to that pace which was necessary and worked while we were in Europe. For my friends who even got the slightly faster accent or *'phonetics',* it has now become a betrayal they carry. Back home, the home-based colleagues are waiting for us to face reality and let the tempo douse down. They wait for us to see that things don't work that way here. Rather than carry a serious face and busy elbow at work, here you need to make calls offline, go to the human resources during office breaks and push for people to remember to do their job. You are almost rejected on both sides; you become a learner of your culture and that of the others.

We all had to find a way to work, adjusting to home while not losing the good side of what we learned abroad. There were times I felt that the journey was too tortuous in the efforts to communicate the message. The message of knowing that only ourselves can harmoniously make the change and straighten our oblique relationship with the foreign lords, driving our career and taking it to its grassroots. Time passed. I waited.

There was a moment of silence.

Then it was time. A time comes when you have no other solutions because no one understands how deep. So you develop a long-term plan. I observed that moment of silence until, like my mentor, I had to extend this hand of fellowship to others, calling on geoscientists and everyone in the oil and gas industry to support a cause for freedom. I founded the iESog platform www.iesog.com to forge this cause for sustenance through our own resources, to reach the heights of development that the other continents enjoy, hoping that one day, this pact with the masses would be understood.

5.5 Compensation for Silence

Before now, academia comprised of respected people and the citadels of learning were places of principles and hope, where the proponents of knowledge and its transference to future generations would not compromise their integrity. Over time, the academia is being ignored; well learned individuals eventually discover that brains alone are not enough in the local petroleum industry.

Their counterparts in the industry who are playing the game of proxy earn more respect and more money. It turns out that, truly, money is not commensurate with hard work.

Some people accuse company labour union branch executives of receiving money, fame, promotion or connection with the authorities, as settlement. Industrial actions are now points of mockery, where our senior friends watching from afar say: "Just give them money and you are over with, they have no ideology." Even the most diligent argue that the local staff are not foolish. They know that this attitude of being easily compensated to keep silent is cheap, but this is all they can get; nothing else. The IOCs will depart as soon as there is no more oil to exploit, or nobody to buy. Through the local union struggle, these workers accept what they can; becoming Caesars who receive what belongs to them. Like the masses, they hope for that free gain from the black gold, coming from only God knows where, in the swamps and far offshore, as long as everyone is 'happy'.

"Oil creates the illusion of a completely changed life, life without work, life for free. Oil is a resource that anaesthetises thought, blurs vision, corrupts." — Ryszard Kapuściński. In your career in the oil and gas industry, you will face this call for blurred vision, for silence, many times. Keeping quiet may seem okay, but takes away what you need to succeed in the transition, the new dispensation of energy sufficiency for host nations.

5.6 Isolation of the Elites

Even where good intentions exist, the isolation of the working elites in oil companies from their immediate neighbours makes them unacceptable to the populace, pushing them to align with external forces. The so called noble mindset that comes with association with the masters, introduced to the recruit through the selection and integration process that sees local personnel working for the multinationals, gradually makes it difficult to remain loyal to their national development interest. Many who could afford luxuries often travel for official and leisure purposes and are then caged up to work for hours behind their screens in secluded and secure bounty towers.

These oil elites who work in the nicely set towers of the oil majors are fully cut off from not only people, physically, but from the emotional attachment of the fatherland. After which they are themselves in need of social therapy, which they cannot get from their inside co-workers; this evolves to a quest for psychological and social-economic needs as palliatives, ever spending and solving problems by just throwing money at them. With time, it is impossible to look back, not alone stand for a cause to better the society. Then there is the effort of survival of these oil professionals to remain relevant in the only system that has brought them solace so far, especially if the alternative developing host countries are failing them in terms of social amenities, and security of life and property.

In addition to the internal severance between a nation and her elites, some Non-governmental Organisations (NGOs) and persons sponsored by foreign powers often use the cover of their platform to find out about the political standing, the weak points of the nation, while collaborations also expose their military strengths, plans and capacities. All these make developing countries easy prey in the hands of their international economic facilitators. If there is fall out, the world powers can target any armouries belonging to their former collaborating countries within hours.

In your pursuit of an oil and gas career, you are standing before nations, ideologies and justice; the role of salvaging your people dangles before you, even at your small workstation. Many local personnel will not use their chances because they never understood the real stakes.

5.7 Sketchy Developmental Programmes

Individual hospitals, individual schools, individual water boreholes, and the basic infrastructures donated by major oil companies, are there as tags. Good as they may be, they do not help a people to have a developmental strategy from a holistic approach. Foreign investors in the oil market are business studs, money shrewd that even when a good intention is set in place as part of social responsibility, personal interests and thoughts about the proceeds get in the way. Those things are not developing Africa; they are helping out the poor who remain poor. Expecting someone else to develop your country is common in Africa and a few other developing parts of the world. Oil workers should not be so gullible as to accept that as true.

The foreign economies emphasise that they are here to see how they can work together with our country; they emphasise interdependence. The local host communities hear that they are coming with social gestures to help or develop their communities for them. Despite years of such social gestures from major oil companies, these have never resulted in coherent overall development. Except the government and the people take developmental issues head on, isolated contributions cannot do the magic. The local oil and gas personnel who are used to flaunting such gestures during signing of memorandums of understanding with host oil communities are tools used to emphasise the shadow while the object itself slips through.

5.8 Absence of Synthesis in Joint Projects

Every major oil and gas company is supposed to play a role in helping the locals grasp the larger picture of the technology. Persons who are supposed to understand the foreign projects end up looking, as if, at a closed four-wall building from either side. Not even the local in-house staff understand the overall projects of major oil companies. What more of the government and monitoring bodies?

The projects are managed from afar. Each local personnel is made to focus on some small units, which is why they may never be able to run the whole project if they were left alone. This is what may happen if the expatriates move back to their own country suddenly, abandoning part of the jointly owned billions of dollar worth oil infrastructures and resource points in the developing country. This emphasises the need to train new oil workers with a different mindset, to fill an increasing gap, despite the seeming decline in the value of oil.

Although there is the need to have a gradual transfer of technology, should the will be there, personnel could be allowed to get a comprehensive developmental plan for over a period of about five years and the rest of the years used for consolidation. To rather release it in bits, over a 30- to 50-year period means you can only get the overall picture at retirement. Only by that time you would have forgotten where you started and the technology would have moved on. That is what is happening in the oil and gas industry. There is piecemeal trainings, sometimes thousands of dollars spent for 3-day punctual training home and abroad, but local careers end up being a matter of position and remuneration. Apprenticeship training is scarce. Cross-posting is more of a compensation for local personnel, many of which are seeking Eldorado.

Some of the persons that left for Europe with me never returned to Nigeria. I came back to do this work of letting you know why we are poor. Why oil money has not developed Nigeria. How we need to start today and build a career that will be useful, diversified and targeted. The comprehensive oil and gas training and transition to newer energy we pioneer (www.iesog.com) takes this into consideration. To help young people find a place of independence, then interdependence among themselves, and to make oil and gas a career like any other. This is an invaluable mindset necessary for a great career at a time of energy transition; the best preparation for the changes that will soon occur in the petroleum industry, a crises that presents unquantifiable opportunities.

5.9 Farm Lords and Farm Hands

I noticed that since I got into the industry, everything was often made ready before I arrived. Whether it was the workstations, the drilling rig, the work documents, everything, even logistics. Some people were often ahead, my employers and their aids, to ensure I only delivered the work in-between. I was to be the farm hand visiting a farm. However, I started to go back to see how those things were put in place and what happened when I left the farm. Today, I would rather start the farm myself. Owning our farm in Nigeria and Africa is what we need to do to succeed as a people. That is the new direction local personnel aiming at a career in the industry should take.

In big oil companies, everything is put in place before workers are made to come in and work in a given section. People achieve a lot without knowing what they are actually doing. The workers may never be able to break free, let alone have their own farm. Proponents of freedom suggest that the only way out is that used by the founding fathers to obtain national independence, which is to form a parallel system, minimise allegiance to the foreign government before attempting to break free. That is after learning the basics in their own schools and development centres at the headquarters abroad. Many Nigerians have followed this path, but who will bell the cat? It requires the ultimate sacrifice, throwing one's life on the line; it is either success or the gallows.

I believe that the battle is first in our minds. If we are free in our minds, we can drive our goals and be bold enough to take the risks of success and independence in the exploitation of our natural resources. We need to understand the mind enslavement approaches; the policies and conditioning that make career profiling after selective recruitment to work. Selective recruitment ensures that only a certain kind of personality is chosen among candidates for employment; career profiling ensures that once these candidates are employed, they are psyched to behave in a predictable way. Let's consider some principles at work when one's career is manipulated: communication is inhibited and self reliance is defeated. This will help to buttress the turning point where someone who should have been smart is enslaved over time, leading to fearful and passive elites who are cold amidst flames in our oil and gas industry community.

Overly Dependent

Making candidates become dependent is an upbringing in the industry. This cannot be overemphasised. You become overly dependent on the oil company system for everything: health care, transport, housing, tax payment, processing of visas, residency, document, family, with your approach to solving each of these changing every time. With regular new solutions and organisation, the dependent employees struggle and can no longer survive on their own. This is done slowly, an upbringing to give you all the focus you need to work. People live for today and forget tomorrow. They will not change if their livelihood is dependent on the system.

While on international assignment, I saw some colleagues rejoice that their tax payment was handled for them, that they never needed to go out of the office to solve any problem. I pitied them because most of them could virtually lose their way when it is suddenly time to go back to their own country; they never learned anything about the way things work in their new place of residency and could benefit nothing from it in any way. Their welfare benefits were outsourced, which they never enjoyed, apart from the monetary equivalent.

Home and abroad, we play second fiddle and make excuses. There might be a wider dependence syndrome which some people call post-slavery or post-colonial trauma. Although the era of slavery and colonisation is over, our people have become used to a fictional master that must hover over everything we do.

Use and Drop

Like the shovel, you may be used when needed and dropped immediately afterwards. In the oil industry, it is easy for personnel to forget that it is their values that matter – not their rights or history. Belonging only matters while you align and support the overall objective of the oil lords. It turns out that some locals who once were involved in the implementation, allocation of duties, organising and convincing their fellows on career matters and setting off one person against the other, under the guidance of the masters, end up being sidelined if they are no longer needed to achieve their role in the oil resources flow to the corporate headquarters.

My mentor was aware of such fiddling with personnel, who were even given offers to entice them and then found them at fault when it was time to sideline them. He once told me that he refused to collect millions of dollars given him in guise of a project, because he saw it was a trap. For example, they might have used him to prove that a local hand was incompetent in that role, where the project was meant to fail, the reason it must be occupied by an expatriate.

Change of Value System

What is termed 'interesting' may be what is completely irrelevant. One is locked in with certain indoctrination of what matters. Our people focus on the hat while the rabbit is introduced below the table by daylight magicians. We are too distracted to care, too opinionated to judge. Our value system has been tampered with over the years and long before Nigeria's independence. We seem to no longer know or care about what matters.

This can be traced back to an older problem, where most people in this part of the world no longer have a single focus. We ask for one thing while working on another. Self-preservation, self-identity and eventually self-esteem have been misplaced. We seem not to know where we come from or where we are going. Our history is so shallow in our minds and anything beyond 200 years ago becomes like a mirage for most historians. The reason for this we already know. Without individual and national values, it is impossible to unite and pursue any cause. We need to reassign our values to what will be binding and a driving force for a better tomorrow. Every person pursuing a career in the oil industry must have a value system to stand the test of time.

Confidentiality Clause

The *confidentiality clause* in most agreements is aimed at making it clear that parties should not divulge information to others not involved in that contract. This should rightly concern information relating to propriety rights. It has now been extended to a way of life, an in-house trend. Perhaps it is the high-tech, or the arduous nature of the work, but oil and gas people have not been too open with others. There is something about the confidentiality clause. A general sphere of secrecy is one of the reasons why the oil and gas industry has lost its connection with the local people. The confidentiality clause has become a general culture in the industry, so much so that as you read this, don't keep it confidential!

Protectionism and aversion for information leaks are why some majors embark on tight office closed networks and email monitoring. In office and out of office, when targeted, it gets to that point where we do not know all that is being monitored. The software, the user logs, telephone tapping; who checks them? Everyone is careful. From confidentiality stems the tracking of individuals, making personnel to live through their work with trepidation.

Ideally, the aims of confidentiality of industry propriety information are:

- To not lose investment gains when results they have endeavoured to obtain after many years of research are given away
- To not release information on the line of next actions, so competitors do not steal the ideas.
- To protect their assets, investigations and trade secrets

In reality:

- Scouted data (not obtained officially or used beyond acceptable agreements) are not to be shown to the outside.
- Today, it is difficult to know who has what among the majors and with respect to the national authorities.
- It is not necessarily the national regulating bodies that have control, as they are often overwhelmed.
- When the authorities are approached for approval, because the national authorities are not expert in everything, they may not know which is which in the data.
- The regulating bodies take too long to respond to requests, as if delay means scrutiny.

The consequences are:

- The rest of the world is not involved in this top confidentiality closure; indigenous firms are ever learning and never having learnt the technology of the oil majors.

- There is shared information among privileged major companies, which remains confidential to others.
- Schools are alienated from industry, especially in developing host nations.
- There is strict indirect control on staff visibility outside the industry, as a way of maintaining the information they have in-house.

While I was studying for my master's degree abroad, visiting staff from major petroleum technical centres came to coach us in seismic interpretation with real seismic images covering major African oil fields. These images ran along many metres of printouts that required several conference tables to display. We had copies to trace on and tie the lines together to understand seismic block interpretation. They told us it was confidential, but it didn't matter that they taught every student and they learnt the trade. Back home in Nigeria, the confidentiality issue is followed as sacrosanct and our students are shown half-page trashed seismic noises for their lesson. Then they are blamed for not being competent when they leave school.

Authorities of many nations, including Nigeria, are gradually lifting the confidentiality clause. The regulating bodies send requests to oil companies to release data to students. This is a good start coming too late, without monitoring of the nature and completeness of the data provided, and its effectiveness. Without monitoring and technical guidance to the person to whom the data is released, there are other consequences. Abuse of data or reduced values due to over circulation of unverified interpretations makes the authorities to further limit confidentiality. Time passes and many students start a career in the petroleum industry clueless as to their actual jobs.

With encouragement from IES Oil and Gas (iESog), experienced oil and gas personnel in Nigeria are beginning to ask for derogation to use their leave time to communicate technical information to schools. Some are doing this after retirement and the idea is the more welcome and supported by concerned sections of the petroleum industry. This can make a change, if the authorisation process for external information communication is simplified for the good of all, albeit in a controlled manner for propriety reasons.

The confidentiality clause serves the benefit of some highly classed big oil players who organise technological information into building blocks handed over to their next generation, while the others in developing host nations get it in bits, in a haphazard manner where they are neither grouped, built upon or consolidated. The data are confidential, which is then understood by the local learner as something not to dig into, something to be kept aside. Ten to twenty years after a well is drilled or abandoned, our students still cannot access the data? No way. Our authorities have to stand by their duty and save our nation. No other person can!

An Unknown

In oil companies and the industry as a whole, there is always a little unknown, there is a not 'fully knowing' maintained. A level of ignorance or mystery of the final truth hangs over everyone and especially over the subjects. Appointments, change in business, sale and acquisition of oil blocks, often happen so abruptly that the local unions often think there is a plot. There is no definition beforehand, as to how things should turn out. Partial ignorance of the highly educated in the oil industry serves as part of the tool of dominance. The younger ones whose careers are still at the pick-up point look on in awe.

They follow blindly until they are overwhelmed by the unknown. Things remain the way they are, even in the face of those who have everything they need to change it. Why? Unknown.

5.10 Reactionary Actions

While in Europe, I noticed that when Nigeria are posted to work in the company headquarters, my French colleagues asked questions about the person before they arrived. They prepared a strategy for each candidate. If he behaves this way, we will follow him that way, if it is the other way, then... They implement policies more than they argue with the people from the subsidiaries. They discuss issues in coffee rooms every morning, agree towards a trend and gradually implement strategic actions without much ado. They avoid issues on the oblique relationships between the headquarters and their subsidiaries in host nations, describing complaints in that direction are petty.

Back home in Nigeria, my people are too busy to even notice when the expatriates arrive. They hope for the better, and when they see the opposite, they suddenly become provoked and act in a rush. They riot or even fight. The rest of the world doesn't take this attitude lightly. They have constructive meetings and programmes to tackle what they describe as "barbaric" action at this end. When talking of relationship with the oil host nations, those personnel who act more reactively in the subsidiary are dealt the worst cards.

The reactive rather than proactive actions at even our top national governance give us a bad name. When everyone reacts at the same time, there is chaos and tension. The instability is a good reason the foreign investors are saying that our country is not favourable for operation; they ask for and take more advantages in major contracts.

We have not been proactive enough to foresee the consequence of what we take for granted, nor do we see the advantages of insights to make sacrifices we should make today for tomorrow. What we ignore in those contracts may provide temporary relief, but also enslave us. When we choose the easy option, like asking the

Operator Company to take produced oil in lieu of our investment share financial contribution in joint ventures, or when we ask the other party to bear all expenditures and we wait for the gain, it shows a weak end. They take advantage of it and throw you the crumbs at their wish.

5.11 Power Tussle

While I was working in Europe, I only knew my head of department (his assistant had no title) and then the heads of a specialist unit of the business. On my return to the Nigerian subsidiary, I was mesmerised by titles, even in the technical section: EGM, DGM, ED, GM, DMD, Chief this or Chief that, Heads of, Coordinators, Supervisors, Managers and Mangers and Managers, most of which had nobody to manage or even a business section. Then you have the EGM cooperate, GM Operations, ED Joint Venture, EGM GSR, DM General Services... the list is infinite.

You may find Executive General Managers or Deputy General Manager and the like in any large companies, but it will be a single office full of responsibilities. In this case, you may have several EGMs and DGMs, in the same division. In fact, some of the managers were unassigned to any department or job. Then I saw everyone tussling for these positions, peers who prided and denigrated those left behind in the tussle. People were ready to give in and give up everything to get there. Personnel who were in the headquarters returned home to take up management positions as a favour. None was focusing on technological transfer.

I hope the new persons aiming to pursue a career in the petroleum industry will not make this mistake. These things come and go in our lifetime, but a mission to give independence and financial freedom to a people blessed with natural resources as we are is a legacy that will live on after you.

Who, then, performed the specialist jobs and tasks here at home? They were either sent abroad or referred to consultants. This rolled on. Despite all the local content bills and monitoring boards pushing for this one major measure to benefit from our oil resources; a quest that has come so far, since the discovery of oil over 60 years ago, we still can't get it right, getting it too late. For many locals who aspired to technical specialisation, it turned out that their work was badly criticised. They were used as guinea pigs to experiment on the dicey technical stuff. Few accepted to take a lower role of semi-specialist, acting as a gateway for the approval of cost-references proposed for technical works to be done elsewhere.

In my case, when I returned to Nigeria, several steps were proposed, contradictory and complicated to follow. I did everything on my part to fast-tracked the processes, letting everyone know I was aware of the strategy to let or hold back technology, irrespective of training. Though I became the first Nigerian to be officially given the position of Geoscientist in the company, I found it was not about me. It was about the rest of the local geosciences community. I discerned that there are things you cannot change from within; some calls require stepping into the wilderness and crossing the Red Sea to a promised land too far to reach. Compromise was not an option.

Tears filled my eyes.

Right there and then, I knew I had to move on, and I was reassured that I was on the right course, with my choices and decisions.

5.12 You Are In For It

Compromised individuals no longer have a choice. Many personnel in the industry wished that they had worked for five to ten years, then left to blow the whistle. As they did not, along the line, they were made to say things they should not have said and sign documents they should not have signed. They became involved. They gave up and hoped that someone else after them would do it – come out and sound the alarm about what went on inside. In *The Overview of the Petroleum Industry*, I went into much detail to explain how the oblique relationship has directed our elites to a particular role, making them no longer play their role in optimising national oil revenues benefits, thereby sinking us into the same problems of poverty and devastation faced by so many dependent oil producing countries.

Despite formal attempts to transfer technology and have the local personnel take charge of the national resources over the years, ancient colonial host countries remain stagnant. While some overly stupefying conspiracy theories exist, there is definitely some element of neo-colonialism in the petroleum industry, which some industry heads and some IOCs recognise, proffering possible honest efforts to start abating it, or at least reduce it to a point where it does not glaringly challenge their positions in host countries. Anti-bribery efforts, springing up in some oil industry majors, are an offspring of this rethink on their part. While efforts are made to curb corruption among oil industry executives, the local personnel should start questioning manipulation of finances at the top, to which they had turned blind eye until now.

We now know that the solution is not a sudden riotous reaction, but to first know how we have been made so feeble. As you plan or already pursue a career in the petroleum industry, you will now learn that to play an important role, you will have to defend yourself from the taming machine of the power game. It has been carried on from time immemorial, in the struggle for control.

Chapter Six

HOW THE SYSTEM WORKS TO TAME YOU

6.1 Everything Works for the System

While the system tames you, everything works for it. Your employers watch your body language with keen interest. They will remember and act on those little side actions of yours. They are witty; they will waste your time if you take them for granted or if you don't read the weak signals. You will become busy with work and money problems. While you are busy chasing the rats, you pulverise the farm land. Never mistake appearance as the real thing, it is often the opposite. Many oil and gas players show humility in guise of their pride. They want your respect; they want you to come to their offices, to see them often, and expose what is on your mind. Senior expatriates think you know what they are thinking; they do their best to gain your support and tame you if they don't.

Along your early career path, your instinct will guide you but often you will behave contrarily because of the external demands and nature of your work. By the time you decide to be objective, you will have been bought over by the status quo. You will be acting in the direction of what every other person does. Below are the attitudes and actions that expose young personnel working in the oil and gas industry, moulding them to a point of surrender, where they give up, remain too busy but aloof and uninvolved in ensuring that oil and gas resources profit the host nation. It can be argued that self awareness is the difference between human and animals. You are tamed when you get to that point of unawareness.

6.2 Emotional Intelligence (EI)

The inability to be in charge of your emotions is why the system is able to tame you. Only with EI do you begin to see that you are literally controlled by your apparent needs, your reactions and your assumptions, sometimes against your true needs. EI soft skills are developed only after several years of working; coming as a new lens through which you begin to see the world, and how small your workplace really is. You discover your past naivety and learn that in your early career, you acted exactly as you were expected to do so, although you thought you were involved in the decision.

Emotional intelligence comes when you do not act or speak out of emotion, and do not overly show it. You learn not to act out of anger or out of 'love', but out of reason, painful for many of us who would always prefer to act out of emotion thinking it is love. This does not mean you should not act out of true love of course. True love calls for sacrifices and doing the right thing, however painful. The emotional guise of love tends to please an immediate feeling.

Many young staff members in the petroleum sector lack EI, the reason they act faster than thinking. Imagine shedding tears or ranting because you are castigated in office. Such a strong display of emotion would not be helpful if it were in a board meeting with mature executives. You look immature if you act in this way, or it could be seen as a pretence and betrayal by the people talking serious business. The ability to remain neutral and act objectively is seen as strength when the stakes are high. If you let your emotions sing, the players up the game, beat you up with wit and tactically subdue you. In your emotions, in your heart, it bites. In some parts of the world, personnel get depressed and even commit suicide, where they live it out alone. In Nigeria and in most parts of Africa, people shout when they address issues, everyone talking at the same time. That noise seems to douse the pain and ignore the bite. But it doesn't remove the taming.

Much information you will receive, many letters you will be sent, comments about you or your work, are all meant to stir up your emotions. As you express yourself, the deciders learn how to predict you, involve or isolate you. Once outside the work circle, of course you may share your emotions or confide in someone. It can get touchy and you are deeply in pain. People are set up, depressed and highly embattled. It is not an easy art to say less, especially when under the radar and with tons of work and societal pressures on your head.

6.3 Word Economy and Selective Action

Word economy helps you to not sell your next line of action. It is not so much about the words you say, but their content. Sometimes people say so many things without saying anything at all, an art mastered by people brought up in royal settings when, for example, they make a speech at a formal occasion such as a feast. Shrewd oil industry executives are economic in what they say, each word counts. A novice is easily detected for wordiness and inconsistency. This is one area where our people need to improve to remain relevant.

When your hierarchy communicates specific things in a few words while paying attention to you, it means that more is lurking around the corner. Don't ignore it when they throw off some light criticisms. Don't retaliate, rather think. They may want you to notice something or to hear more from you. In practice, never denigrate the head or your hierarchy at such times. They would be listening out for the slightest reaction. Choose when and what to communicate, not letting loose because you are prompted.

Select which fights to undertake, not launching randomly. Don't join a fight you do not understand, you may end up only being used. Once you select a fight, make sure it is worth it and follow it up to the end. Avoid open arguments. The fight is within, saliently and sometimes silently. Ensure you emphasise the objective behind your proposal, the intention and not the person involved. Though you know that people push one another around by their actions, it should always be about issues, not those people, when talking. Once you are economic with your words, it is assumed you are action prone and reserved enough to manage challenges.

6.4 Your Predictability

I see young recruits lay out their expectations, their plans, their visions and hopes, bare before their observers, who unbeknown to them are their competitors. The senior friends laugh inwardly. They have mastered the art of predictability while feigning neutrality. They say a word and wait for you to complete it. In that part of the world, people who have big dreams do not display them too early. Else they may be antagonised or schemed off.

Certain persons set traps so we will react or fail in some ways. Even when you outline to one manager what plan the other manager has for you, you may just find that the new manager wants to change it to be in control; or to prove the other manager wrong. It is not about you, but your predictability. When you react against a poor situation someone wishes to place you in, they strike by knowing it matters to you. Their insinuations begin to look real. You need to learn to jump the hurdles, by reducing your predictability. Know that those who stand before you are themselves asking questions.

Don't be too specific in your definitions. Don't bare your requests to those who can't grant them. As soon as you want something, every other person around you desires the same. Colleagues go to managers and protest about one person getting a benefit if they did not. Truly, there may be times when you are not given a promotion or an opportunity because you are not clearly defined, but being overly defined in your early arrival is also a weakness. It is good to define yourself only as it applies to the expected opportunity at the time. You may temporarily lose some bait. In the long run, however, you save yourself from hard targets heading towards specifics.

Isolation Tactics

In my other book, *Overview of the Petroleum Industry*, I mentioned that there are two sides to the table in the conference room. This side and that side; side A and side B. The oil and gas farm lords want your side to be clearly defined. They want you to choose a side and have a clear-cut location. They want to know exactly where to locate you. Your fortress. They start asking you several questions on first meeting. They watch who you associate with and classify you. It is natural to ask and receive answers to questions like: Where are you from? When were you recruited? Where do you live? Married, how many children? What are your plans? With these answers, even a child can set you up. No problem in good times; in bad times, however, they will play on your weaknesses, that soft spot, that one thing that will tune you as a guitar.

Isolation, for positive or negative reasons, is dangerous. It makes you traceable and your destruction would not be shared by anyone. There are times when going on cross-posting is just not suitable for you. There are times when becoming a manager or being offered a top singled-out position is not good for you. There are times when being in charge of that big investment money is not the best thing for you. Anything that singles you out should be seen as a challenge, a temptation, a trust to guard and defend. Knowing these things already helps you and prepares you for taking your career to the next level, when you take the offer.

You may not refuse or deny when you are singled out for a good opportunity. Take it with reservation, being positive but watching out. To him that much is given, much is required. The more you are singled out, the more conspicuous you become. Offline, quickly associate and mingle with those on both sides; that way, even with a two-headed arrow, no one will know the direction to shoot across the fence.

When You Get Carried Away

Long-lasting winners never take their fight beyond their goal. They watch their own success with apprehension, knowing that they will lose what they do not keep. They know that time and chance play a role and that one success does not determine the other. The Bible says that he who thinks he stands, let him take heed so that he does not fall. The naive go-getter gets carried away. With one success he predicts and announces the near future to all and sundry. Peers sometimes think that your win is their loss. As soon as you win, the loser plans a comeback. Those who hear your celebration wonder if anything could have stopped you. The future successes you have announced generate many obstacles and some people will act to the contrary to find out what happens in the event that you lose.

Never over-assume; over time situations will work to get you where they want you to be. You may, in the short term, get carried away, thinking you have arrived. You miss the point and become overly dependent.

Platonic Relationships

The amateur local staff are unaware of relationships without benefits. Their senior counterparts consolidated in the system have learnt the lessons of lip smiles, of shallow friendships and spy relationships. They meet, show an interest, listen intently, and leave with few answers. They could be your 'friend' for years without any actual sentiment. On the flipside, Nigeria is a contagious society where every relationship expects more. It turns out to be a good virtue that serves a rather exploitative purpose on oil industry workers by their experienced counterparts. The amateur is open, eager to associate and to express himself. He gives it all; he is surprised and feels betrayed when his experienced counterparts don't go all the way with him when trouble arises. His counterparts never saw it the same way; they never intended for close association of oneness in purpose and intent. It was just an ancient inheritance of wisdom, tenacity, control, playing out to get feedback from you.

It takes a long time for many in the industry to understand that the office relationships among staff and with company management are detached all along and that he who goes in head and toes gets sucked up. No one was really committed to you; forget the parties and the close-up concerns about family and health; it was business all along. It was just interests, not that the office was your second home. The insiders played a game to get the rules across and learn to win. They told you something flimsy about someone just so you would tell them something serious about someone else. They revealed nothing so you would reveal something. Though there are always some honest friends at all levels, many personnel unfortunately sideline them for those with 'silver linen'.

6.5 Good Image and Influence

Things should always look okay. Everyone needs to show a sign of happiness. There is a natural tendency of the society to aflame in an oily environment, so oil and gas workers are expected to ignore individual differences and the many troubles. Personnel allow one person to talk and listen before the other. It is civilised and calm. It works and everything keeps going on as usual. Few young recruits learn to fit in fast; they only add a comment to the thread making the rounds in a coffee room. Most new threads are introduced by the boss and it is he who sets the tone. The over-zealous new staff member may want to analyse problems, raise issues or criticise something. Poor gesture; he may tagged as unrefined. Here people know that words and complaints don't change reality. When a group of local staff members shout and argue and talk of the bad news in their society, the listeners on the other side nod ardently to this self bastardisation and image destruction.

Career growth in the industry entails body language that reassures everyone. The idea of leadership is that you do not disturb already troubled waters. Show the good image. You may need to laugh when nothing is funny, as long as it is a comment by a witty top shot. You then get into a mood where no one complains. You become immune to the TV critics of the industry, to the poverty around you, to the trouble in politics, to society. In fact politics is prohibited in communication and on the intranets or online platforms of oil companies. It is a top level life of luxury, affluence and everyone sees only good. They flaunt it, just like top politicians and government ministers who are sometimes afloat, even when the house and nation is burning beneath their feet. That is how the core of that local employee in the petroleum industry was sold.

One way to show a good image is to use an alibi. The most difficult sanctions against local differing personnel are not announced by those visibly at the top or directly in charge. Rather, other persons, such as line supervisors or a gentle proxy in human resources are the front-line soldiers. They are the contact persons, the pointmen. Be careful when that top player who was against you remain quiet and pretend that all is well. There might be something cooking which will soon hit you from out of nowhere. The person used to hit you might not even know that he is just a tool. The farm lords like to remain in the shadows when dealing with an erring staff member. When you are hit, you will only remember that you were warned.

Putting forward a good image means that anything that attracts a scandal is strongly repressed. If you have blown the whistle, you can no longer benefit from the team whose wrong doing you announce. They are on the chase, you have to be on the run or pay the price.

Options and Choice

The idea that every objective question set during exams has at least one answer is something that has been ingrained in our memory since school times. When you had to choose one of many options, doing all to get that one which was right. Many leave school still carrying this, believing that there is an absolutely right answer and an absolutely wrong answer and that they are included in the options presented to them. During oil company career meetings, you are often presented with options. You hear colleagues whispering, troubleshooting and brainstorming to know what to choose. Unknown to many, there may be no real options but the same proposal in different shades. There may be no one right answer or no wrong answer. In real life, there could be many right options, and different routes to the same destinations.

The young employee thinks that he chooses, or thinks he is involved in the decision. Deciders ask you what you think, what you want, what if... At the end is just what you have been designed to achieve. Options can sometimes be a snare, in which case it is like a flow chart in a computer program with an 'if else loop'. The programmer knows each outcome and has directed it to the end purpose of the program. Either way, the choices you have not set by yourself have definite outcomes and those outcomes have been put in the equation towards the desired answer. It is even worse when the options you are given are for comfort items. The only time you really choose is when you create the options.

Creating a Dependency Link

Comfort and control are synonymous. Young local staff members think that once they have comfort, they are freer to live their lives as they want. As mentioned earlier, the problem is that comforts come at a price, especially comforts that were given to you. Why those guys on the streets are so free is because they have no comforts. Every object of comfort is a weight which limits you. An invisible rope ties each comfort to a control structure. In one case, I was going for an international assignment and was offered a large cargo quota by the company. I was expected to carry all my household items and even local food provisions that would serve me for years. There was hope of another quota to bring them back. Though it was free and to my advantage, I rejected the offer. I travelled free. With such an opportunity, some personnel move all their household effects, including the log of wood they may never burn.

To be free as you carry on with your career, do only things you need to do, rather than do what you don't need to do just because there is an offer. A long-term dependency link is why you are tamed.

6.6 Money and Time

In all capitalistic economies and in the politics of control, it is the balance between time and money that oils the wheels of the cart. In this call for technological transfer from the big oil corporation's headquarters to developing countries, I notice that the more your time is taken up, the less involved you become. People are sent on a long-term developmental cross-posting or training, just so that they miss out on an opportunity for which they are due. Take an analogy in the aviation industry; if you were to become a captain in a fixed winged airplane, but then sent to start learning to fly a chopper – what a tactic? Going to learn how to fly a rocket into space is even a bigger distraction. No matter the material benefits, training coming too late and too slow is useless. Our local elites in the oil industry often miss out on being the captain, then eventually leave the chopper to go for the rocket. Your time is being bought for few benefits. You never get into space.

The persons intended to truly go into space will learn flying everything comprehensively in few months, then focus on the main objective. The learning skills, reports and development you have made while learning slowly to fly and taking in all the flaws are harnessed and given to the actual persons who will spend even less time learning to man a rocket. You are just not in charge and your time is stage-managed. You have paid for the flight training but get nothing at the end. If not by cash, then by your resources which will be used to fuel the plane, chopper and rocket at all stages.

While promising you get some interests and training, you are to provide the capital for the chopper or the rocket and the funding to travel into outer space, which you don't control. These costs would be recovered from you, before you get to the end of the journey. You never get to outer space because only your time and resources were needed. Visibly, no harm meant; it's just business by the farm lords over the farmland and the farm hands.

6.7 Avoid Shadows of Failure

While those on horseback may fall, they never associate to the fall. They do everything to make it look as if all is under control. Falling is necessary to ride well; they redefine the meaning of falling. Your senior colleagues analyse their work and indirectly call for support, making reference to experts and specialists whose validation they earn. They buy everyone over to avoid being burnt. It's more of image than reality. The idea of not getting it right is scary.

In some parts of the world, many work to not be on the downside. If eventually they find themselves there, too bad. They remain with those of their own clan. Attempts to come out of the dungeon are subjugated. If you struggle and succeed in coming out, people exclaim at the miracle. Few may wonder if everything is alright and may give you a slight down push to be sure you are really up. If you are a visible failure, nobody wants to side with you. A major part of oil and gas players avoid the unlucky, as no justification of the reason for the bad luck makes it desirable. Perhaps this is why in some quarters someone picks up guns and shoots everyone around them!

Our own people share their complaints; how unlucky they are, how they are maltreated, thinking it attracts pity. People take time to tell you why they are a failure, as if that changes the situation. The thinking that reasons justify lack of success is probably why that is an option for many. Local personnel in oil companies provide the yardstick by which they are judged. That very low price tag you presented is what you will be labelled with. This label keeps you from your rights and responsi-bilities. We should spend more time to learn, do the right thing and work our way out of failures, rather than wallow on the low side. It's about what works, not reasons.

6.8 Official Generosity

Probably because of the blame on the role of fossil fuel on environmental degradation, the quest for a good image in the industry is high. The good image is necessary, both inside and outside. Daily posts on the company intranet need to reassure the workers of what the company is doing. The workers in any case have less real contact with society and are happy to share in the game of goodness. They do punctual visible acts of kindness to oil host communities, put them up on TV and newspapers and perhaps feel good about it. The community sometimes feels taken for granted and complains. What the oil companies pay as ransom in cases of vandalism threats to oil installations is more than what they pay at will to appease the community. To put it another way, some of the gestures are more to secure peace rather than charity.

Some of the majors consider that they should not owe the community by contractual obligations, rather it should be what they do at will to a society which must be appeased, in which it is difficult to do business; a country where the government has not played its role and has left the environment and people in shambles. The people themselves are confused, not knowing which way to go. There is a stake on who to blame. Blaming the wrong person means you get no compensation. If a host community is overly compensated, other parts of the nation complain of bias. If not treated well, others complain of the insensitive nature of the oil and gas business.

Since the oil companies will be criticised any way, they do only what is perceived as clearly 'good' gestures. This show of generosity by acts of kindness should open the way for them. Because honesty and true intentions are hard to use as pleasing acts, the showmanship serves as a replacement. As these official generous acts are announced, many are appeased and calm.

6.9 Focus on Target

Even when dealing with the opponents, the oil majors take it one at a time. One target, layer by layer, with dedication until completion. Dealing with many facets at the same time could weaken the strength of the gun. Imagine you were to shoot ten lions at a time. You rather choose them one after the other when they are isolated. You make friends with the other nine lions temporarily.

Every effort is made by personnel not to be the focus of a vendetta. Team members are overly subservient at heart, but are positive and show bravery at meetings. Everyone works in the direction of a target. Each focus may have several aspects, but everyone works until that is achieved. There is an objective at all levels of operation, global, in-country, branch or department. Each appraisal is focussed, with an A, B, or C plan to be achieved at specified times. In this single-minded focus, the individual loses self-awareness. Every counter-current is dicey; when everybody is going that way, you too do. You don't know you have been restrained, like a pet, to behave the way you do. You know the direction to go; you surely don't know the destination.

Work from the End to the Beginning

Before any project, the oil masters see the end. The focus is that end. Sometimes, the rest of the team see the way, but never sees the end until it is completed. That is why they get there. They plan all the way through; the reason you need to read the weak signals. Every action stems from an end result. People say that the present determines the future, but remember that in the oil industry it often seems the future decides the present.

If the Managing Director of a multinational tells you that "if the company branch of the labour union continues to ask for more the way they do, the company may no longer be sustainable," you have to stop and think. That sentence could mean that the company will sell its assets and leave. A good local labour union thinks it through, weighs its options and decides if what it is asking for is their legitimate workable benefits.

If the company decides to start recruiting contractors, it could mean that they are boycotting the in-house staff system where the empowered personnel ask for accountability and control. Contractors do not have such power and as such, the headquarters run the system directly, even when increasing local employees as contractors. The company argues that local content policies demanding more recruitment of Nigerians have been applied; not showing anybody that control by the locals is reducing. The company holds all its facilities as rented properties. Accommodation, office space, rigs and vehicles, could all be rented. It seems costlier at face value, but it allows for easier mobility. The company may sell major projects or hold only a small percentage. They gradually become technical advisers to fields where they own shares instead of full operators. They reduce investment in exploration and focus on producing fields. These are all signs that they are ready to abandon.

If all employees are contractors and all assets out-sourced, then walking away in the event of an unfavourable decision by a government or a local host nation's elite becomes easy. They expect that you will see the minimal commitment, in a wait-and-see attitude. That "the company may no longer be sustainable" means the company may opt to leave. All options go in the direction of the plan, to stay or to leave, as the corporate company was ready all along. There is already a 60- to 100-year plan on how to exploit oil and gas and leave, from the first day the oil majors set their feet on the oil host nation's soil.

At every stage of major oil company activities, there are plans to the end. Planning to the end helps manoeuvrability and maintains control by leaving all options open. In corporate business strategising at the top level, the local executives might not see it coming. They scream and wail as if everything was an accident. They do not learn to plan or see existing plans to the end. The new indigenous oil and gas candidates, to get into the industry, must develop this foresight and be trained to see the future. They need to emplace the end result in the future and work towards it.

Changing the Rule of the Game

We respect authorities in our cultures, even revere them. We allow the patriarchs to tell us that we got it right or wrong. Those who work on the other side of the divide know that it is better to have robots assess them than to leave it to a subjective mind. Our senior partners count more on machines rather than on humans. Not that humans are not better as programmers of the machines. It is the judgement they doubt, not their ability. They fear that their destiny is at stake when determined by other humans who turn out to be silent competitors.

In other parts of the world, it is fatal to count on somebody whose interest may be opposing to vet you. They like written agreements so the rules don't change halfway. When you think it is not clear, that you may have an explanation to escape, you are wrong. They did not forget the rules; you have been trapped. The rules become clear at the end, often in your disinterest. Not being conscious of this as a young local employee could be detrimental, especially if you are seen to be going too fast and too wisely. The masters want you to know it is not just about the job, the skills and the work objectives. It is about learning to work together, under the same interests that had existed long before you were recruited.

When your equals on that side of the table where the masters sit cannot get it right, they change the rules. Though they said the sun was white, they explain that it is white in shades of yellow, or too bright for anyone to confirm the colour. You may find out only after drilling that the intended oil well, which you warned about, was drilled just for policy reasons, or that the well being dry was a good thing so you let go the block. Once you get that trophy, without their support, they will affirm that trophy is now useless. Attention focuses on who was the more cheerful, who got there without effort. Explanations will fiddle out your achievements. Some would actually disdain what you carry.

Your career attainments will not come directly from meeting your end mark on the race track. It is more of evolving rules of the tournament as determined by the allies, as only with their support can they declare you a winner.

Achieve Easy but Forecast Challenge

The watch word in the industry is 'challenge'. Right from recruitment everyone wants you to mention that word. How you like challenges, how you will meet challenges. Then for each project, each piece of technical work, whenever objective is set, anything that matters, is a challenge. It keeps everyone on guard. It makes anything, coming up, look mighty and people feel satisfied that they are working impressively. The error is to carry it across after it has been completed. Once the objective is achieved, the language changes. Nothing in the past is a challenge. Now the word is 'easy', 'simple' and 'common'. It doesn't matter how hard it was for you to get there, it was easy.

At this point, it is a mistake to emphasise, directly, that what has been achieved was difficult. The art is to say how easy it was, even over-emphasise the ease so that others will now see how able you are. However, a novice recruit may be disheartened as s/he hears that such mighty achievements are easy. Making an achievement easy is one way they keep you at bay and always on edge.

A work done over years could be presented in one go by an old staff and leave you dumbfounded. Just take your time and it will be alright. Once you have mastered the basics and learnt from the masters, it is time to become one. The solution is to set your own goals, your own targets, and your own approach. Not that it will be easy, but that gives you a chance. Take the objectives set by your boss and adapt them to your style. It may take time, but it is better than attempting to depend on the validation of those on the other side, which may not come quickly. Instead, that attempt to be like them overshadows you and hinders you from bringing out your best.

The Winning Victim

In the image game, between major oil corporations and the rest of the world, there is the need to keep a low profile while taking the big lead. Have you wondered why the IOCs are often peaceably negotiating, planning and patiently running their affairs with their partners? It is about achieving the objective and not the show of force. While the amateur believes in verbal complaints, the masters sometimes accept defeat as a tactic. In our immediate society, people fight to the last man, dying rather than making peace. The top oil and gas company lords take all the downside and get what they want, while protecting what they have.

At war, you may need to temporarily accept defeat so as to re-strategise and claim victory next time, justly, fairly and deservedly. The onlookers will be on your side. Playing the victim is a strategy they use to ensure the next time it will be their turn to be promoted. During negotiations, the lords play hard and soft, tough and easy, taking one and leaving the other. They don't attempt to get it all at once. When dealing with governments, they navigate in-between, not too loyal or too dismissive.

The lack of this strategy is why in developing countries people fight a war they are sure to lose, hoping that some international help will come. A different approach, which works on the oil industry career path, is to quietly aim for the important things, what you can surely get, even if it seems you are trailing at all-time low. A feeling of being victimised makes everyone allow you to be rewarded. Henceforth, take note when somebody plays a loser while he or she wins. Being a victim or prejudiced is not the same as being a failure; victimisation is tied to an external control, while you are blamed for being a loser. This requires their kind of insight, as there are good and bad times to be a victim or a winner, an art mastered by only a few. Remember never to get carried away.

6.10 Catching Attention

What did you do this weekend? Where did you spend your vacation? Those were common questions while I was doing my cultural immersion, living with a French family and when I was in the language school in Royan, south-western France. The questions followed me to school when I started my post-graduate studies in Pau, in southern France. My answer had to be interesting; one needed to do or think out a good answer. Doing something interesting and being interesting, having wit, was an art you needed to learn early in France.

Once done with your first answer, you are asked 'what you will do this evening'. You are told of an interesting cinema, game, whatever, that will be held by the other team. I was learning to fish for interesting things, to know things, so you have got to be prepared. You must remain interesting. It's a test of your creativity and vitality. Walking slowly, talking as if there is nothing much happening, is a loss. It is a sign you are giving up.

When in a group, people act in an excited fashion, appraise common talk, and laugh intermittently as if something interesting was actually said. As mentioned before, that is especially if the person saying it is a boss who should naturally be witty and interesting. As a young recruit pursuing a career in this milieu, you too have to be there. You need to laugh with others and whine when it is whining time. If everybody is laughing at the same time, saying the same things about the same game, finding everything interesting in their pointed direction, they have likely been tamed. They have been made to think of the minor and forget the major. They swing left and right like a pendulum and you can always predict the next direction.

6.11 Novelty and Royalty

The expatriates I see in the workplace are not directly asking to be respected, but they show it. They show that they have the heritage of civilisation. They carry a sense of superiority. Only 'royal persons' with noble qualities and endowment, in this case technological knowhow, will carry such air of novelty. The look, to mystify – occasional dramatic icing, follow their expressions. Come to think of it, the Western oil headquarters have achieved much for their nations since the beginning of history. Their people should be proud of that. Should you visit those countries, you will see modern railways, ancient castles, modern airports, ancient roads, all beautiful and made through the contribution of colonies and self effort.

That is the point of comparison, where one comes from. If one has better roots, a supposed heritage, that novelty and special privileges give one an edge. Then a form of boldness learnt from childhood is used to crown it all, making others seem little. At the point of amazement, they now sell to you the way, the programme that will bring you up to speed. They send you on a wild goose chase, different periods of incubation to make you worthy, when you most needed to be involved in the now.

You could do all the trainings, all the works, yet get back to where you started to find that the conditions have changed. In my own experience, I learned under and with the masters, and had the masters learn under me. I worked at the headquarters and at the other affiliates, onshore and offshore. Did I ever get there? That is a question I have not been able to answer because my 'there' has evolved as well. Never expecting or wanting to rule with the system, I choose to side with our people and drive a cause while time permits. Will novelty ever be bought? Is it for everyone? Maybe yes, maybe no – I choose the assignment that is before me; the mission of helping many who want to pursue a career in the energy industry to know the truth and play their role in developing oil host nations. I choose the cross on this side than choose the throne on the other side. With that, I rest my case.

6.12 The Discredited Option

You will be treated in two groups; one wins while the other loses. The person that loses is given the finishing crush, so the other learns the consequences. That one person who was judged to get it right melts every critic when the loser is being punished. It confirms that the masters have good judgement, judging some to be good.

Options always presents as a good and a bad choice. Winning not only has to do with getting it right on the right job. Character and loyalty will be assessed. When the erring person is punished, everyone gets the salient message: toe the line or get the worst treatment. The penalty is meted out by your own people, so that it is all seems decided by them. The lords then maintain a clean hand, a pacifying role as arbitrators.

Even if everything is not clear in your workplace, there is an obvious warning sound; a 'be careful' feeling, which works as when a domestic animal pet is tamed. That fear, that something hanging over like a hood, covers everyone's eyes. You see personnel en masse behave as a herd, as sheep without shepherd. You now understand why over many years, the status quo is maintained. Even when the oil and gas elites visit those countries, they come back unchanged and remain in the circle perfectly tamed and controlled. The quest of independence in our oil and gas resources generation and management becomes ever distant.

6.13 Your Worries when Lonely

As we come towards the end of the taming principle, it is once again important to revisit the issue of emotions and worry. The battle, they say, is in the mind. You show it, and they stir you up to see how you will behave. It is time to control and apply that aspect of your intelligence.

The fear of isolation, loneliness amidst people who would have been your own, is why many in the industry remain attached to the system. For some oil personnel, detachment from the workplace is now stronger than detachment from family. Even at retirement, hearts fail because of a sense of partial disconnection with the office. The very work they loathed, when they were stressed-out during mid-career crises, they find they cannot leave. When your hierarchy frowns or withdraws from you at work, it is painful because as you were recruited by the company and integrated into the team, you got a feeling of family. You were cared for, placed in good hotels with good salaries. You have discussions at the workplace, wear nice cloths, and have a feeling of the world rocking around you. Then, suddenly, there is the silence, the occasional scrutiny and mixed feelings.

Your hierarchy goes this way and that, watching how you react to the best offers and to the worst announcements of possible ugly happenings. If you have gone too far, then you no longer share your emotions with friends and loved ones, but with colleagues in the office arena or club. At these times, your peers may already be feeling the same, in a fixed career game that is not all smooth sailing. Many end up boosting their survival by waiting for that one thing that spices up their life, which is the month end 'alert'. The salary and the benefits continue until the personnel finds out that money doesn't just solve it all.

Then comes the most challenging vicious circle, where the worries of the past show in the present and that evidence becomes the gloom that prevents future success. Your emotion shows and anyone can prick it. Once you have shown those peaks of emotion, then you can be tamed to the desired end. How you react to 'too-good to be-true' promises matter. This is like the right and left turn on a steering wheel, used to steer you to any direction. The masters do not give their emotions away. They disappoint you when you least expect. Like not coming to dinner, especially if it is a personal invitation. They see a catch in it. It is like throwing a hook with bait for the fish. They take the big steps in their desired direction.

6.14 Awaiting Set Hopes

In every career meeting, in all communications, there is always a promise, no matter how flimsy. It is like a pacifier. No problem about meeting the promise. In the next career meeting, they can give excuses, change the plan, but always promise another. A situation where some local staff do not see the workplace as a place where they contribute but a place where salary is received, anything can be believed. The hopes of Eldorado, be it physical or emotional, makes many give up every sense of moral duty while on the journey. They focus on the end, not the means. Our people live by hope, so it works when top executives keeps talking of the next plan for them, the next thing; a way of playing to their needs and desires, to keep them going, while running the affairs of the day.

While I was in France, I noticed that people asked you '*comment?*', that is '*how?*'. The journey was as important as the destination. Any hope of future glory that does not reflect in the present is challenged or looked upon with suspicion. They fathom that if the process is faulty, the beautiful end is either a bait or not real. In our immediate society, it is some past examples and future hopes that keep everyone in the system. They do anything, hoping to secure their future in the company, at least, to make ends meet and keep other loose ends tied. Whilst local personnel are being rewarded punctually, they lose the bigger picture of a holistic success that should affect the very tenets of society.

At the end of the taming process, we all work directly or indirectly for the system of resource flow and control, enshrined in the principle of the 'orange and the knife', elaborated in my other book, *Overview of the Petroleum Industry*.

Chapter Seven

DEVELOPING OIL PRODUCING NATIONS

7.1 Causes of Underdevelopment

There is no doubt that oil and gas industry enriches every nation, including Nigeria. A lot of development would not have been possible without it. The accompanying negative effects – the deterioration and poverty, which come with oil and gas resources, make many begin to wonder if oil has been a curse rather than a blessing.

The neglect of the upstream has relegated the on-shore and offshore oil and gas technology to a distanced euhemerised process that the majority have little knowledge of. So oil wealth is seen as a blessing, a manna that is not worked for. Politicians wait to receive rent and allocations that fuel a society where merit no longer counts and wealth is not seen as something obtained by effort. This concept is the engine that has fuelled corruption and made the elites not accountable to the people. Corruption has always been at the top of the list of our problems, but many forget to look at why there is corruption in the first place.

It turns out that the problem of corruption should be at the bottom of the list, to be understood only if we understand the causes. Why do the very people who are supposed to know, to make a change, the employed oil and gas elites, sometimes work against their own country? The reason is not just the obvious better organisational and financial benefits that the IOCs tend to offer, but some more salient motives. I learnt about these during the time I was given the opportunity to have a deeper understanding, through the unique scholarship and work training scheme pioneered by my mentor, the late Dr Kingsley Ojoh. He was transited to glory on Friday the 23rd of April 2018. May his gentle soul rest in perfect peace. I am happy to have published the first book on this matter a year earlier in 2017, *Overview of the Petroleum Industry*, dedicated to him.

I hope that, as you consider these factors, you too will gain the insight on why developing nations, and their youth, lose sight of their focus once they mingle with their successful counterparts in the West, within a multinational company setting. This explains the wanton response of opportune indigenous persons in reaction to their degrading social, economic and natural environment, succumbing to external influences and growing dependence. The factors below are both individual and collective attitudes and actions, which also explains the entrenchment of corruption in our society.

Broken Bridges

The first thing the major oil companies do, once they select and employ a young innocent candidate, is to isolate him from his past. They sanctify him to the heart of the company's mission. They do this by changing his setting, language and scope; like sending him on a training programme in a remote location. He is introduced to a new language, people and tasks. He becomes suddenly engrossed, having to let go of all he had before.

In those days, if there was no mobile phone or internet where he was located, he was lost for a while. Even in these days of sophisticated communication, he may find himself on a site without internet network, where he can only communicate rarely with his loved ones. He has burned his bridges behind him. This is the beginning of a long journey that sometimes ends in nemesis. After many years, you don't establish lasting human relationships except with life partners or family members who have had to adapt to a missing spouse or parent. Due to prioritising your work, you have long lost all affiliations with loved ones at home, who now see and communicate with you as the one without interest or attachment. The network you build with people while in cross-posting is also lost, except with those related to your current work where you serve as a gateway for your country's affiliate office.

While in Europe, people ask you when you are going to return to develop your home country. There is something ominous in that question. Once you leave Europe, as with many others, often the tie is lost, and you are now back home, away in West Africa, Nigeria, where less than 5% of the emails you send out will be taken serious thereafter. Your merely returning to Africa cuts off a major part of the relationships you had with the rest of the world.

Fear of Own People

Ken Saro-Wiwa wrote that Africa kills her sun. Perhaps I see some truth in that, where we are not protected by our own people. Antagonism overshadows the very bright sun, which is mocked when it becomes a small flicker of light. There is a fear by the young recruit that any failure in the oil company will be mocked by the people s/he left behind, and that even the oil company will use our own people to frustrate any attempt to succeed outside their grasp. Some people ascribe this to diverse tribe and ethnic make-up, while others believe it was installed by the division among the people wrought by colonisation, where successful allies of the colonial masters were seen as traitors. Whatever the cause, the consequences are obvious.

Nigerians travel far and wide, as if running away from home. Most successful persons in our cultures don't return to their hometown or village for fear of being 'killed' by their clanspeople. It boils down to the concept of the stranger being an angel. Our young recruits are happy to find an escape. They dive in head-over-heels into this new career pursuit, never considering that there may be a need for a rethink in the future. Hopefully, they will realise that things are changing and everyone will need to have a home. Like Nelson Mandela said, "Never forget where you come from." In our days, no one cares about you that much in the company. Everyone is finding their way; with recent changes in the industry, no one feels safe.

Fear of Extinction

With limited options available in a challenging society, the new oil and gas employee holds on to the one opportunity he has. Your relations often say, "Nigeria is risky, hold on to your job in the multinational." Instability in our country is one of the reasons we become so dependent. Without health, safety and basic infrastructure, any marriage that is proposed to our youth is good enough. There is the fear that once you are not on the side of the multinationals, that is, on the side of the oil majors who are the major investors, no one will patronise you and your objectives will die a natural death.

Things are changing though; nations are waking up to the truth. China is questioning the dollar as the currency for international reserve; nationals are questioning the role of foreign oil exploration companies as leaders instead of contractors in developing countries. The internet has made people communicate more. Individuals now travel worldwide and are beginning to think for themselves. Conscience is creeping in on the slaves as well as on the masters. It is time to go all out and be ready to live life to the full, or die.

Time Value Difference

When I worked in Europe, I could wake up at 8.00 am and be early to work by 8.30 am. In Lagos, Nigeria, I woke up by 5.30 am and sometimes after harsh traffic I was late for the resumption time of 7.15 am. The majority rent houses in the highly chic neighbourhood close to the office to meet up on time. The money they spend in a year could solve half the problems of their village communities, but now they end up broke every month. Then comes the time to fuel the generator or to stay out because of the heat and then to attend to uncountable solicitations of help and religious programmes and parties.

In addition to the time constraints instilled in us by the absence of orderly development and population growth, is that demanded by the multinationals who expect us to be cool-headed and face our work the way it is done in their own country. Most of the expatriates live in hotels or special residences where everything is taken care of for them. Most leave their spouse behind in their own country and are here on regular rotations, having all the time in the world. Competing for presence at work with them is like suicide to the local staff. These local staff members keep quiet, putting in so much to maintain their career in the international petroleum industry work setting established in the towers of Port Harcourt, Warri, Lagos and Abuja, among other oil cities.

Personnel take out more loans to personally provide classy enabling amenities to keep up, thereby having to work even harder to pay off these debts. They become overly loyal to get promotion and the grants that come with it. Who is still talking of conscience at such a time? You wonder why you have not heard that a top oil executive in Nigeria resigned because of a policy contrary to what he believed in. When you get to that point where you feel constrained, you need time to plan an exit strategy from the current burden and find a new source of livelihood. The oil business culture makes you lack just that. Often, by the time you work it out, discovering that you need to be free, it is too late. Your hands have been soiled and your needs set to obligations that call on you to maintain your involvements. The old are afraid and the young too young to be ready.

The Human Factor

There is a human factor everywhere that tends towards error. In Africa we need to learn to deal with it as is done elsewhere. Else it becomes an African factor. Individual and collective responsibility towards good is non-negotiable. The Bible says that it was since the time of old that the wickedness of man was great on the surface of the earth, such that every intent of his heart was continually evil. Why don't most Africans get this? Being good is not really innate in man; it is either you learn that it pays to be good or you are forced to be by law. When humans are involved, thing don't go the good way naturally; failing to plan is planning to fail. The approach where we expect things to change themselves does not work. Nature seems to give you the bad leftover if you do not go for the good.

A collective effort towards improvement produces synergy. It suffices to say that many problems we face as individuals could have been avoided if we had prevented a similar problem when it happened to someone else. What goes around comes around. In Nigeria, individuals tend to think that if something doesn't affect them right away, then they should not care. In the labour market for example, the lack of a collective career plan, is why many are struggling. A more caring plan could be to work as if protecting a heritage for the unborn generation.

You need to be a decider of your own destiny and that of the duty entrusted in your hands towards your nation. It is more a long-term career plan, which should be in harmony with good governance, for us and for our children. Learn that it is not necessarily the visible collective actions that count, but the personal actions directed towards a goal, sown as a seed that germinates and grows in the future for everyone to harvest. Each generation serving as pacesetter to build capacity for the next, works together to limit policy failings, which has cost us so much as a people.

Absence of Practical Training

Without practical training, young people leave university to live on autopilot, a life without creative achievements. oWhat I see all around are so called empowerment promoters who gather young people for their personal rewards; motivational talks abound from people who have no success themselves. They give verbal encouragement instead of practical training.

After one of my awareness talks in NYSC camps, the organisers kindly suggested that I use an open hall in the local government secretariats for my training. Many other organisations share the same space, showing slides and displaying posters to young graduates. That was okay for an awareness session, but not for a training pro-gramme. I found they gave little in terms of actual capacity development. On our part, we have rather set up a workshop and a laboratory for the iESog training, where people will obtain energy generating skills. We include food processing in our agricultural biomass section, so that no trainee is left behind. We advise the trainees that as they do oil and gas, they should also have a farm in their backyard. We ensure that everyone who comes into our facility learns a trade, something to do, once finishing the programme. That is true capacity building at the grassroots.

If our government or political leaders were doing the same, there would be national centres for skills acquisition, many of them for the oil and gas metier that has served in sustaining this country for over half a century. The ministers of petroleum would have built the biggest energy technology centre in Nigeria. Our technical experts would, in addition to mastering technologies that harness oil and gas in a local and prosperous way, be able to drive the transition to renewable energy and create core and ancillary jobs for millions in Nigeria. Without human capacity, no matter what we budget and spend on infrastructure, they cannot be properly utilised if our own people cannot install and manage them. That is why it costs so much to do so little. Politicians would rather party and hold electioneering conventions than build human capacity to run technological centres. Requests of practicable decentralised power grids, modular refineries, state policing and all the smaller facets of solutions to local problems fuelling the national crises, are pushed away, perhaps for fear of empowering one section of the country more than the other.

In President Muhammadu Buhari's electioneering slogan, he said that corruption will kill Nigeria unless Nigerians kill corruption. I find that the fear of disunity will kill Nigeria unless Nigeria kills the fear of disunity. It is this fear that the country will divide that dissembles meritocratic process, where persons who show the tendency of maintaining the status quo are prioritised over efficiency. It seems many Nigerians are afraid of voting in any president who will raise critical issues on the polity. They vote for you if they think you can hold the country together. When the person does not perform well on other grounds, they say they have no other choice but to keep "managing". They only vote the person out if they see that things are falling apart, not because the country is not moving ahead. In any case they are not sure their vote counts.

That is why there is no regional policy, local electrical grids, economic restructuring and state rights to mineral resources. Since the federal government collects all the proceeds from the resources and not just taxes, the allocation from the centre gives the impression that money comes from politics. The problem is that at federal level, we lack what we also lack in the federation units. We lack practical centres that reach the whole nation, as central facilities do not reach each region. This is why our heritage is shaky – we do the same thing for years but never realise that it may never work that way.

Resources without Capacity

Personal and national resources without capacity lead to crises. Not giving attention to building human capacity, followed by technological capacity, is why the oil and gas sector has rather made us dependent on foreign aid. It is the capacity to exploit the resources that makes you king.

In Nigeria, a lot of effort is placed on how to collect oil revenues to fund the national budget, which essentially runs the recurrent expenditure. If you use all your resources to pay salaries or daily provisions, or to pay others for basic infrastructure, you are not moving ahead. Then a government that plans a budget with infrastructural development such as roads and bridges is acclaimed. Yet that is not the point. If you pay foreigners to do your road, the money leaves your country. If you had paid a local civil engineering firm, the money will circulate within the country. They will patronise other endeavours within the nation, boosting the local economy. Besides, they will be able to maintain the roads and it will cost less.

The only true form of empowerment is when that budget is used for capacity building, to generate the funds – such as to explore and produce our own oil, and build our energy grids. This is the same solution necessary to build factories and manufacturing plants. We need to think in terms of creative capacity and not in terms of the ability to buy goods. The new mindset proposed in this book is that of oil and gas career persons who seek to produce rather than seek for pay. In the food chain, the most important process that sustains life is at the base of the pyramid, not at the apex.

It is the same at the personal level, as it is at the national. Many think that because you work in the oil industry you have money to spare. The truth is, you get money, but have nothing to spare. You spend at the level of the income and life class, buy cars and houses as other colleagues, and throw money to problems you do not have time to solve. Once entangled, you can no longer start and grow your own investment. An oil business is big, not something you can start with peanuts. So you are stuck dreaming and never waking up. Same thing happens to our country; that we have oil does not mean we have the capacity to produce it. Those who produce it are in control. Resources without capacity are no resources at all. If everyone in the oil company is a manager, what will they manage? Yet many oil and gas professionals in our country, copying their political counterparts that are at the helm of national affairs, strive for that position to oversee how money is spent and not the generation of it.

Frustration and Apathy

The word 'frustrated' is common in some multinational oil companies and you will marvel at the attention given to it. You hear something like, "He was frustrated" or "You will be frustrated." You don't want to be. Job success, security and comfort imply alienation from stressful environs, the challenging streets with heavy traffic, security issues, health issues and poverty. You have almost succeeded in avoiding these by staying in your well air-conditioned office and working with only your computer and a few colleagues. Any frustration within will be dicey, as you will be caught up between the devil and the blue sea. You are no longer prepared to cope with the outside world. You are too scared, not willing to lose the relative peace.

The apathy that results from this phobia calls for advisers, most of which turn out to be poor advisers. They are in a similar situation with you and can only see from inside the box. Most of your work friends are in the mood to visibly please and those who don't antagonise you are equally not in the right position to advise you on certain issues. Advising you to jump a ditch, which turns out rather to be a mountain, is an example of how dreams can be different from reality. Confusion comes with multiple pieces of advice from inexperienced peers who themselves may be using you to test out their own fears. In the midst of this, personnel stick to what they know how to do best – remain in their comfort zone, to retain their source of daily bread and allegiance to the paymaster.

Not Keeping it Private

There is a known rule of success that there are things you don't discuss with everybody. It's true that when you don't tell others about your escapades while they shout out theirs, you look foolish. Then when pride catches up with you, and you share the goal, you may lose it. Sharing your plans with your colleagues in the office is one way to kill it, not because they wish to, but because you are looking for help from people who also need help. While your colleagues are good chatting mates for general social life issues, discussing how you plan to go for non-immediate-profit goals, or do some societal duties, or break free to build a patriotic empire, will be like sharing the content of a wedding with monks.

There is a slight antagonism against striking personal achievements among peers, such that even those who would be happy for you work against you. Steps of indirection are taken to stop you. The best way to share your dreams is to achieve them. *Show, don't tell* is a phrase every good novelist knows. Either due to human envy or the general fear of change, people rather maintain continuity than allow you to advance beyond the common boundary. Maintaining continuity is the struggle against the inevitable. Some people will move. Eventually things will change, but will come finding many unprepared. As they say, he who does not plan to succeed plans to fail. Again, without initiative, you live what life has left behind.

Having been bragging all along, most personnel in the pursuit of their career get broken, unsteady and fearful because of the shame when things do not turn out as expected. When people don't get what they want, they tend to think that thing is hard to get; they make it so.

Due to such experiences, they not only fear to evolve, but like the proverbial Pharisee, they stop you too from moving. As you share your goals and dreams with those who cannot pursue them with you, they picture the reasons they have not embarked on the journey by themselves, and share it with you. If those reasons stop them, it might as well stop you. With all that information and counter information, you are overwhelmed. A sense of something ominous hovers over one's sphere, not letting you take in charge of your decisions and life.

Lack of Patriotism

If many Nigerians find themselves beyond the shores of this nation, they will not return, no matter what challenges they face in the foreign land. They are ready to pay the price there, than return to this country. It is not just about the suffering, but the diminishing sense of hope, a feeling of a lie they find lurking behind the national structure and policies of government, which affects the inherent believe in a fatherland. They feel that we do not face our problems and any efforts to make amends are suppressed.

Those in government are not helping matters – they ignore citizens' demands. A popular call like economic restructuring of the nation is swept under the carpet. The central government prefers to collect national proceeds and share, rather than provide an enabling environment where competitive businesses flourish and pay taxes. There is seemingly nothing else to hold us together strongly and drive the passion to fight for one's fatherland. In the midst of all of this, politicians wield power and comfort. A "no one cares" attitude leads to people seeking easier ways to meet their needs. It leads to jungle justice, bribery and corruption.

Some people believe that this is a hangover, a consequence of the destabilisation of ancient traditional ties that were upheld in local communities in Africa. The allegiance of the indigenous allies to the colonial masters, as against the divide with the larger populace, made the common man to see them as betrayers. Then came the new political class, stemming from this unwarranted roots, the upheavals of the first national *coup d'état* leading to the assassination of some founding fathers of Nigeria, and eventually to civil war.

It seems that we have moved in outward configuration, but inwardly and tactically, the Nigerian problem still persists. A problem that finds its way in the oil and gas sector, disuniting the cause which otherwise would have brought our elites together to foster a common goal. The career paths of the Igbo, Yoruba and Hausa and the other national tribes in the country can easily be mesmerised by surprises when anyone introduces tribal sentiment in appointments. The oil farm lords see how our emotions are heightening and allow that to serve their purpose. We do the job, then fight our neighbour for their own gain. Once people are fighting, you can appease them and get them to follow you.

You are not better than your homeland, no matter your individual success. The oil business involves so much money, and those who run it are often desperate to get to their goal; it becomes a duty – it is indeed a pressing duty, to generate energy, but not by any means. This results in some of the worst scandals in the 'hostile' developing host nations. In these scandals, the local employees have not helped to protect their national resources. This we must change as a new generation of solution providers in the local energy sector.

Corrupted by the Job

Corruption is real in the industry, as well as at employee level. According to the Global Economic Crime Survey of 2014, misappropriation of asset and procurement related fraud will pose the highest risks to the companies in this generation. In the face of corrupt regulating authorities and bureaucrats, some oil companies overstep their boundaries with lobbying. The desires that drive more ambitions often drive fraud in a society where the checks are low.

Such drives include increasing:
- Quest for material gains
- Attempt at cost reduction
- Desire to acquire oil assets
- Degrading moral standards
- Loss of the sense of integrity
- Pressure to yield results
- Competitive growth and requirements
- Strive to top company positions

The purchase of goods and services by company staff and their contractors has remained one of the loose ends. They justify their actions by believing that the system itself is corrupt, at higher levels, and that they are only taking a small part. There are personnel who believe that once they have obtained short-term corrupt gains, they will leave the company or industry to pursue a good cause that will return these benefits to outer society. That deceit to oneself often leads to procrastination of their intended future. Other personnel feel that the rest of the world is not as hardworking. A wanton feeling of shared guilt leads to poor ethical policies in most companies and monitoring authorities.

The regulators have failed to:
- Effect internal controls
- Trust personnel based only on objective verification
- Purge the larger society of coronal fraud
- Create a transparent environment where fraud would be easily detected
- Penalise employees cheating on behalf of the employer

Sometimes check systems are not regular, making them ineffective. Without using the checklist like a pilot on every flight, there is no way to trust the engines. Hoping that things will just work out, when there are no repetitive checks and active actions, is why figurative leaders soon lose grip of national sanctity, despite their initial body language against corruption. Rules and regulations seem not to work in a society where corruption and rot are fostered by law enforcement agencies such as the police and the judiciary themselves.

Local young start-ups are too eager to succeed, too afraid of major loss in a hardly supportive investment environment, as such, too willing to partner with the big names, that they soon become compromised. The solution would be to reduce greed, to diversify sources of benefits and to have joy in less material gains; to value other benefits of freedom and a free mind. Don't start to compete if you can't stand the turbulent waves.

Although corruption is worldwide, flagrantly tolerating it in our society has debased national standards and discouraged those honest personnel who would have stood for what they believe. Such honest people are also not encouraged if a role in whistle blowing will rather expose them and leave them unprotected in the hands of the undertakers. As President Barrack Obama said in his address to the African Union in July of 2015, corruption is not only a problem of Africa but also of those who do business from Africa. They siphon billions of dollars worth of resources and kickbacks from countries who do not have enough to cater for the basic needs of the people.

While achieving their work behind the scene, increasing technical knowhow makes some staff to become like computer robots, digital wise but street averse. The yearly appraisal in the office is inversely proportional to street smartness. It really takes a lot to not lose one's integrity, even with just office politics. Imagine what goes on with politicians at national levels.

As you pursue a career in the industry, do not get carried away and lose your thoughts on integrity. Don't be overshadowed by the normal trends of a day that passes with the wind. Personality and individual original inclination matter. What do you spend time on? There is a dream that is like a call. You have to weigh the obligation versus the dream. You need on one hand to provide for your family and on the other hand achieve what you believe in. Don't do a job for life just because of what you will earn. Choose what you like to do, no matter the cost. If you do a high-risk job you like, it may or may not kill you. But if you don't like your job, no matter which one, it will kill you slowly. The solution is to find out where your obligations meet your dreams, that point where it is safe to take the risk!

7.2 Solutions for Developing Oil Nations

The first solution was to ask why and look at the causes of underdevelopment. The answers to the problem are the opposite of the causes. Among the solutions will be to have true national leaders and followers, who work for the long term, leading and producing leaders. We need to learn to add value to our resources, then move from an awareness situation to valuing persons with insider knowledge of what goes on in the oil industry. Most of them are working for the IOCs and they need to care. While working to adopt the right mindset, we need to know that the current crop of politicians in Nigeria are not professionals; they know how to spend money but not how to generate it.

A culture of honesty and a feel for fairplay from our leaders, a sense of belonging to a nation we all love and attempt to build, are paramount. What we have today are people in government who enrich themselves from national coffers with resources generated by oil rents. They work to increase and cart away wealth rather than for the people. This situation, known as kleptocracy, works in a semi-democratic government like ours, because the details of oil export are not understood by the masses. If the masses generated the produce individually in their backyards, they would be more sensitive to taxation. It is only in military or ancient oligarchies that you can easily subject a people like this without they reacting. Unfortunately we have been rendered docile, decimated by ethnic and religious sentiments, such that we cannot galvanise efforts to demand what we want, to stop the nefarious government of the day by the use of industrial actions or public civil disobedience. We have not learnt to hold our leaders accountable, and as if to ransom, until they do the needful.

If the people ask for the implementation of a national consensus on the polity, then let the political elites lead others to give it a hearing. If they ask for economic decentralisation, then let it be considered. Let people have a sense of being given attention. We cannot make real progress until we work for a common good, break our problems into bits and solve them in packets. Individuals pursuing a career in the industry will be able to contribute if local control units of oil drilling, refining, power generation and mineral mining are authorised. That is when our young recruits in oil and gas would not have the only option of working for the gigantic IOCs in a gigantic refinery for a gigantic output for national grid. Making everything big is why they remain insurmountable.

At the national level, there is the need to foster positive changes for individuals and collectively, using our oil and gas heritage through an energy mix, to a developed country where we will have the basic infrastructures and sustainable energy solutions. Without unity and a sense of something to protect, our elites in the industry will work against their own national interest.

Some of the factors to consider are:
- Derivative processing
- National leadership and followership
- Fatherland awareness programme

- Insider capacity building
- Mindset change culture
- Honesty and accountability

Until people come together as one, they do not achieve great things. This has been the bane of African nationals and their counterparts in developed nations. I encourage persons seeking a career in the Nigerian oil sector to consider this truth; the young persons who want to delve into the upstream as employees or independent workers will need to work together. Two united persons are stronger than a thousand persons in disarray.

7.3 Derivative Processing – the Fifth Dimension

Those aiming to get into today the industry should adopt a see-through approach to all their pursuits and work engagements. They should see through installations, investments and technological activities. They should look beyond the gain to the diversification and enhancement of their operations. I propose a new value added concept of *derivative processing* where you should use the gold you mine, use the oil you produce, learn the tenets of power sources to its generation as well as its use. We must follow through to generate and monitor the energy footprints in our own environment. This is not only applicable to the petroleum industry. Farmers should learn to process and use the crops they produce; herdsmen and ranchers should process and sell their milk themselves.

Value-added or *derivative geoscience* is a solution where geosciences personnel should take their basic mining skills to the utilisation of what they produce. I call this the *Fifth dimension* or *5D*, adding value through utility, to mining or harnessing of resources. Henceforth everyone working in the energy industry should adopt a complete participation in the energy chain, from the punctual identification of resource in length or depth (1D), through the area or surface (2D), through the volume or quantity (3D), over the time of production

(4D), and then use the actual resource to add value (5D). How much our thinking and approach enhances our collective development cannot be overemphasised. Fifth dimension, as illustrated below, should become the watch word.

1D   >   2D >   3D >   4D >   5D

Length   Area   Volume   Time   Value

While engineers and other metiers in the oil and gas industry focus on equipment and installations, the geoscience professionals interact more with Mother Earth, as some call it. It is therefore the geosciences personnel that should care more and lead the harnessing and management, as well as good use of the resources. They should strive to replenish what they harness, and ensure it is not done beyond the natural quota of life replenishment. To do this, they need to know how to process the fossil fuel resources and minerals from ore stage to final product. Then go beyond that, by using those ores or metals for engineering projects. The engineering geologist, geotechnical personnel and mineral miners must become much more involved in utility.

The keyword in derivative processing is value. If the silicon layers and metal conductors used to produce solar panels come from the geologists mining efforts, then let the geologist produce the solar panels. If the Sun is part of the universal planetary space object, then let the geologist understand its use in energy generation. The base resource for the wind turbines and almost everything used to generate energy gadgets is produced by the geoscientists. Oil and gas and plastics are also by-products of petroleum in its crude state. Let's take the responsibility of using even the waste from these to generate energy and recycle them back to nature.

The idea of value added geoscience can be applied to avert the major mistakes that ruin oil producing areas. In the early days of oil and gas development, our people focussed on allowing the production and selling of the petroleum crude without being involved and taking charge of that production and its processing locally. They left out what was important, the technology and diversification of the utility-based plants, such as those necessary for refining crude oil and for power generation. Let us be involved and use what we produce to change our nation.

EP&R not E&P – Definition Matters

The best example of where the lack of 5D, lack of derivative thinking, has done us great harm is in missing the mastery and continually being effective in refining oil resources. We were deceived by calling oil and gas activities: *Exploration and Production (E&P)*. It should rightly be *Exploration, Production and Refining* (EP&R). Because refining was not emphasized, the few young geosciences and engineering students studied and focused on exploration and production. Our national government treated refining as if it were something on the side. E&P technology was fundamental, but refining was king.

Without refining, there was no way to put the value of petrol to use in our own society. Having some implanted refineries was not enough, if we did not focus on doing it yourself. Not domiciling the derivation of the petrochemicals and the utility of the refined products is even a worse failing. Refining therefore should have been a major course in universities and a regular mandatory course in all petroleum training and studies Nigerians undergo. The consequences of this negligence are glaring. Subsequent PMS importation, subsidy, and dysfunctional refineries show the policy failure.

We can clarify it further to mean exploration, production, refining and usage – in power, petrochemicals and other utilities. Over 50% of what you see around you is made of petroleum products, at least as one of its components. We allowed the exploitation upstream and are busy discussing how to manage the funds in the downstream, not having the capacity to independently produce and refine petroleum as a country. We behave as if what mattered was the resource; actually it was the technology for both producing and refining. It was the ability to go from crude oil to actual benefits of the crude in society.

7.4 Leadership and Followership

This will not be overemphasised, as we all know the challenges of leadership in our country. The problems in Nigeria are multifaceted; it becomes difficult to do a reform when the root cause is structural and hierarchical. The Nigerian story is like that of a young man who did not make plans for the future and maintained the same income he had as a bachelor while spending and feeling comfortable. Now he is married with children; with increase in the number of family members, he realises that the situation is not sustainable. He needs to double his efforts, cutting back on his excesses on one hand and having to earn double of what he had to be able to survive. Had he planned for more income, bigger house, stable power supply, accessible infrastructure, early enough, life would have been easier even with a larger family.

The story can also be likened to a young student who approached his first year in the university without paying much attention to his grade point average, and then failed most of his courses. For the rest of his university years, more course modules came up while he still needed to carry over past courses he failed. If he had passed the earlier courses, he would have had enough time to solve future problems as they come.

Nigeria's population has more than tripled since the early years after her independence. The problem we have today is no longer that of how to generate electricity, but how to raise the persons, create the mindset, build the tools, develop the competence, harness the resources, manage the finances, provide the leadership, and more, that are needed to generate that electricity. If we had those basic tenets of developments alongside our growth, generating electricity would be simpler. To solve the human, infrastructural and economic problems in Nigeria today, we need to double, if not more than triple our efforts. The radical measures, which such efforts call for, are not possible with the recycling of incompetent, non-visionary old leaders who continue to fail in policy making, instead of directing the efforts to solve the problem at its roots.

Perhaps the lack of followership results from lack of leadership. We need selfless leaders who will work for the yet unborn. In every sector, every region and every career type, we need great minds to pave the way for others. Leadership makes everyone to look at an apex, not by constraints, but by shining the light that guides the way. It brings about harmony. It takes one person to build a few disciplines and eventually as many disciplines of the disciples. This is where good manning at the top is important for the development of every nation. I seize this opportunity to again remember my late mentor, Dr Kinsley Ojoh, for the good steps he took for many, recruiting us in campuses across Nigeria and following up our progress. He did all this so that we too may carry on with the struggle for truth in the oil and gas industry.

In the last one-on-one meeting I had with Dr Kinsley Ojoh about a year before he passed away, I asked him salient questions. I was concerned about the challenges related to the nature of our work, operator-partner-authority relationship, how one's integrity is put to the test. Among the answers was the resolve to maintain a line of people who will always lead the next generation to do good in the milieu of the oil and gas companies. He emphasised honesty and integrity. He told me to always be concerned about whatever was within my control. He told me I should start where I am, not minding what stage or position I found myself in the industry. He said my actions would define me and any consequences could only steer me towards my goal. I should be ready to give up all to find my way. It was a turning point; I left, ready to face and live life.

Living life to the full was the advice he had followed himself; he was always with his wife whenever he travelled, not waiting for his retirement before going for that special vacation. He sponsored programmes not limited to oil and gas business, but including charity, supportive outreaches and ministry of the Church. He travelled by road to remote university towns to carry out indigenous recruitments and scholarship programmes on campuses, what few executives in oil companies would dare attempt.

Leadership requires sacrifice and the strong will to wade through the storm. My mentor was persecuted and challenged abroad and at home, yet he rose in his career to become a director. Then, business politics got so high, that at a time he told me that you needed to be strong and spiritual just to stay alive. He was a devout Christian; I came into his office many times at lunch time to find that he was fasting. In the schemes of things and as God will have it, Dr Kingsley Ojoh's hard work, and his ability to survive challenges, has endeared him to many. His highly intelligent and astute oil business knowledge took him to an executive position in the company. He passed away shortly after his retirement as Executive Director, Special Adviser and Ethics Officer of Total E&P Nigeria Limited.

How many of those who flag around as leaders are so deep? If you get the same opportunity, in the pursuit of your career in the petroleum industry, will you do the same? It was at this point I decided that I would move a step further, away from top office politics, to a more societal role where I can provide guidance for the young ones. In the early career days of Dr Kingsley Ojoh, there was no internet and no social network. The national regulating authorities knew less and there were not the thousands of graduates returning to Nigeria from some of the best petroleum schools in the world. Today we have the upper hand to carry the message further. We need to start where he stopped. We have to emerge wholly after an incubation period, like our Lord and Saviour Jesus Christ who always told his followers to keep it calm until he launched out fully at once. The Saviour knew the world will attempt to stop him too early. We learn to make our success gradational before it is fully blown, so it is not denied or avoided.

Great leaders see the future, change the rules and change the world. Dr Kingsley Ojoh attempted to change the rules, in capacity building. Instead of going for cross-posting when personnel have grey hair, he worked hard to see that young Nigerian geosciences recruits would go on international assignments while early enough. For a few of us, it was first as independent scholars, before we joined the IOCs. He insisted on exchange, the same as when the persons from other countries came to Nigeria for an operational base, to learn rig operations. He hoped we would learn how things worked at international levels early enough, then apply the lessons back home before we were too old.

During my PhD in Sciences of the Universe, I met researchers studying space and time. We were planetary sciences students and calculated distances between spatial bodies and tried to solve equations in nine dimensions. I met people from different nations during our professional meetings and we discussed why regions of Africa remained so poor, despite their resources and highly educated ones among us even then. We discussed my country, philosophised over the why, who to blame and who not. Capacity was built in me. A lot was expected of me, both within and outside the institution. I promised not to fail.

I told those around me that it would be disappointing if I kept quiet and did nothing on my return. It would be a disappointment to the good people in TotalFinaElf,

NNPC and the French Government. They gave me a chance to participate in the *Development of Talent Scholarship Award* and history will question how it was possible to remain docile after such exposure and capacity building. I promised to dismember the machineries which made my people not stand up for what they believed. Forces tried to stop me, but I had to stand up to play my role.

7.5 Fatherland Awareness Programmes

Many Nigerians are fed by foreign TV programmes. Our schools teach children Western culture. Good as such exposure may be, we gradually lose the coherency that makes a nation. There is the need for the actions of the government, indigenous community, and local programmes in our agencies of socialisation to be stronger than the "I don't care" feeling. The government needs to show Nigerians that their lives matter as individuals, and reawaken their consciences. We must root indigenous national programmes at our tribal and cultural units.

Young recruits in the oil industry are indirectly warned that they will be lynched if they pose the slightest threat to the oil and gas system. Many believe that our government is desperate and will do anything, including sacrificing a Nigerian, to ensure that oil money is not lost for even one day. What is our government saying or doing? What is it showing to counter this accusation? In France for example, the consciousness of their country seems to run in their veins. Important national issues are discussed in informal settings, disseminating national trends. Visiting persons during such discussions won't understand the details, not only because of the language, but special semantics with which such matters are discussed to show inner and citizenry rights and patterns. In Nigeria, we should work towards such cohesion at all levels. Our government leaders should come out into the open consistently and reassure their people.

They should:

- Remind indigenous staff of our national history, sacrifices of heroes, roles of the first local engineers and implications of their actions for their future and that of their children.

- Make these workers feel relevant. Remove the "I don't care". Condemn evil in society and eliminate tribal and religious bigotry in politics.

- Provide security and health to everyone and show fairness and equality, promoting and giving recognition for acts of collaboration between elites and society.

Only then will young persons seeking a career in the industry feel safe and justify the cause to fight for any national cause outside the one proposed by their paymaster. Those days, I faced many people who reminded me that if it were to be for my own government alone, I would not have obtained the international sponsorship. Today, international sponsorships from our government are available. Such has to be made to become an 'insider programme' where studying is rather treated as work training, with integrated capacity building within the relevant industry. Then let the government carry out popular campaigns to let the people become aware of the impact that this is making in the society.

7.6 Insider Capacity Building

Capacity is everything. Do not accept capacity building by looking from a distance. Be involved in the actual operations of the oil field. Developmental programmes by our governments should be done in collaboration with the oil companies; the national authorities should integrate indigenous firms in in-house trainings of the operators of their joint ventures. Our universities should have an official exchange programme with the oil companies registered in Nigeria. The regulatory bodies should follow it up and proffer more insider scholarships, insider training, and insider operating and investing for the future.

In the quest for technological transfer, it should no longer be between "you" and "I", sitting on different sides of the bench, but moving on to sit on the same bench with our foreign investors. This may raise initial eyebrows, but in the long run it is the only way to go. It is like deciding when going to drive a car, if by sitting inside it scares you. You need to get in, you will learn. No remote control can do it better. Remotely operating the oil and gas resources from afar is a bane to Nigerian utilisation of their resources. When oil was discovered in Nigeria, we should have started from the inside, but due to our limited awareness then, we started from outside to manage the national oil and gas exploration and production. Till today, we are participating from the outside. Participating from the inside means being the one operating the oil and gas fields, only employing experts or expatriates where necessary. Let us face the technical challenges, even make mistakes and learn.

7.7 Mindset Change Culture

We arrived in Europe at the onset of our post-graduate programme with a mindset that the people we met would be willing to help us, to give us more support, most importantly ignore insufficiencies if there were any. We felt we got the opposite; while there was support in terms of highlighting the standards, even kind words, we were made to go through the hardest way, to mitigate any existing lacuna. We took additional courses for good reasons, attempted final year mathematics courses since we were Master's geosciences students; we learned in a different language and our papers were marked more stringently. Our handwriting and poor French language capabilities betrayed us. At a point it was as if it was important to prove that we could not be good enough to cope. They might have been nice, just an effort to not validate us wrongly. They call it *"apposer le tampon"*, a kind of "rubber-stamping" to approve the candidate. Some of my colleagues fell along the line.

Later in the company office, I found the same traits; we needed to prove ourselves even more. Every new person or new idea was first challenged until proven trustworthy. This approach produced good and bad outcomes. Good outcomes if we survived, bad outcomes for those who failed. A lesson I learnt, living in European society, learning the meticulousness of the French, is that nothing was to go uncircumspect. Nothing goes for nothing; no detail is negligible. We can learn from this; the way of thinking which demands that everything is questioned. We need to know that the other person is not here to do you good, but merely to do business. This frame of mind, which we must adopt, does not call for violence, unnecessary argument or being difficult. You may even be more polite, but still knowing how to play the game.

7.8 Honesty and Transparency

It is one thing to be honest and another to be accountable. As a people we need to be accountable, guided by law and order. Good wishes are not enough; there is a paperwork to do, as part of due diligence, in the execution of our professional and civic roles. When the rule of law is emplaced concretely over time, it becomes a culture and the next generation grows to have only one option that they see – that of transparency in legality. With transparency comes proven honesty, then accountability. Probity becomes necessary when onlookers see through.

Many IOCs and NOCs today are worried about possible allegations of corruption in execution of contracts, which may jeopardise their position in this era of intense checks. It has become necessary to integrate a culture of integrity in the training and development of industry personnel. On fraud preventive measures, highlighting consequences would ensure that young staff make progress in the right direction to secure the image of their company, the nation and the industry at large. On the side of the authorities, it is not enough to wish, they have to take institutional steps. On a lighter note, why not for example give lectures and print documents on the importance of corruption-free careers to petroleum industry recruits, as a must-read pamphlet on entry?

Gone are the days when companies grow because they are openly corrupt. The idea that you should not be transparent, so that the oil industry can patronise you, is becoming obsolete. Whistle blowers of the new generation will track you down and place it on the internet and social network leaks. There are factors mentioned below that are moral issues, others which are a matter of documentation and responsibilities of the parties involved, all of which when not undertaken lead to corruption.

Employer's responsibilities:
- Educate staff in the anti-fraud policy laws of their country.
- Emphasise cross-supervision between staff and management, where employees participate in decision making.
- Bring down the understanding of debit notes and accounting adjustments to the level of the technical hands.
- Encourage cross-functionality, not allowing a situation where a member of staff avoids the processing of a transaction by any other colleague to conceal existing corrupt practices.

Staff responsibilities:
- Have a global understanding of the overall system of your company activity, despite division of labour.

- Avoid non-professional and exclusive relationship with customers or suppliers, which may lead to conflict of interests in decision making.
- Show extra boldness and challenge superiors or clients where they have links with the deciding person in the contract and procurement section.
- Report abuse, such as boss intimidation or appraising you poorly to silence you during shady contractual matters.

Responsibilities of the authorities:
- Control unjustified significant financial inflow into accounts of persons in positions of contracting or decision making.
- Unravel unusual, irrational or inconsistent behaviour and sudden change in career movement to hide corruption in the system.
- Check people living big and money vanishing, while their declarable and taxable earning or investments do not justify it.
- Emplace traceability in business transactions, identification and proofs for record purposes, and revisit such documentation regularly.

Internet banking and a cashless society are helping in Nigeria, but the limited availability of surveillance cameras and use of amateur hidden cameras is still setting us back. A capture of bribe collection should be posted online when caught and top level meetings should be video recorded. With the internet age and with the social media, one advantage is that evil may no longer be hidden for long. Incompetent leaders are soon exposed when they commit blunders, while Nigerian online geeks add comics and cartoons that spread such inefficiencies like a virus. For a country striving to raise its head amidst suppression by old oligarchies, we must take up every option, and the career paths in the energy industry must provide avenues for participation of their to-be elites.

There must be continuing interest in identifying fraud as the interest in developing new fraud techniques is increasing. Every suspicion must be reported and investigated. A cause for concern is the insiders who serve as the gateway for collusive corrupt practices, a situation that is hard to track by companies. The over-socialisation leading to compromise and the attractiveness of profit, remain a high tempting standard in a poor inequitable society. These things must be taken care of as we launch into energy diversification and renewable energy investment, so that it is not corrupted along the same lines.

Chapter Eight

THE ENERGY INDUSTRY
– A CHANGING CONCERN

As the world sounds the prepare-to-abandon alarm on petroleum resources, developing oil host nations must (like the ship masters) muster the most competent men to get the best of the life boats and ride safely to the energy future – rather than jump to sea.

8.1 Interests Win

In 2006, I walked into one of the world's top oil and gas technical centres in Europe for a six-month Master's internship programme. I met the deputy manager, the technical admin head, of the Nigerian affiliate support unit in the centre. He took a few minutes to acquaint himself with the fact that I was on a scholarship and that after this internship, I would proceed to do a PhD at the university. Right there and then he decided that I would not get a free scholarship; that is, I would have to work for the three years spent on my PhD to compensate for any contribution the company was making towards my sponsorship.

A scholarship stipend was not a salary, either in France or in Nigeria. How this would work out; how I would share my time between the centre and my university was an issue that was resolved after a few months. Everybody turned a blind eye, since the arrangement would provide a win-win situation for both parties. I was coming from West Africa, there were no clear ties, and communication with sponsors at the subsidiary was poor. We had no choice, so to say. For the company, I would be accomplishing some work tasks; for my university, there were scientific objectives I must achieve. For me, working and studying was an opportunity to gain experience.

It is important to note that the processes leading to this scholarship took over two years to complete in Nigeria, after which selected graduating students prepared for months for this lifetime opportunity to be freely sponsored. We were at the Consulate in Lagos and Embassy in Abuja. It appeared in bulletins as free good training gestures of the IOCs and Joint Venture in Nigeria for locals. It was a no-bond scholarship stating that I had no other obligations to the company. There was a fanfare in our schools at graduation and even the welcome by CNOUS (the French administrative body for scholars) at the Charles de Gaulle airport in Paris was great. Then we settled, started studying and living in the society. The Masters programme was over and here I was, now having to indirectly pay for my sponsorship by secondment to the company.

This position of doctoral attachment to the company, three years working internship, revealed some aspects of an oil industry career to me. It turned out that what is not written on paper held sway. I was at the edge, shifting from opportunities to isolation. I was treated as an insider, as well as an outsider. As usual there were the progressives and the regressives, the good and the bad, even the ugly. The arrangement worked for the interest of all parties – no blame.

There are a lot of things which are not clear, for the interest of several parties, in the oil and gas industry. Interests win. Being too clear in your expectations could obscure some clarity. Getting into the industry is an objective to many who aspire to earn a good living. This good ambition, albeit a poor objective, has deterred candidates from seeing their actual roles when they do get in. Indigenous oil workers find only later in their careers that they should not have been poor in the first place, or so desperate as they sought to gain employment.

The above anecdote of my internship is to buttress a point in the ongoing argument in the energy sector. This argument of interests comes late in the timing of technological transfer, making us to ever be in pursuit of what is next. Let's see why getting in and out of the oil industry should be easier than it is and why you have a role to play in that sector, passive or active. Let's see how the current call for an immediate exit from oil and gas, and the move into a new energy future, though a necessity, is directed by such interests. What is your own interest?

8.2 The Argument

Is Crude Oil in the Past?

Is coal in the past? Just like felling wood is in the past? Fishing sharks is in the past? There are arguments for and against, but the actual argument is on how to maintain energy sufficiency while keeping our planet clean. Each nation's interest affects this debate. A country like Nigeria is forecast to grow in population from 200 million to 500 million in a few years' time. That is an increasing energy requirement, meaning that even if fossil fuel will provide only 30% of our energy then (from the about 90% it is today), petroleum production will rather be increased to reach sufficiency. It may take more time to be done with fossil fuel in oil producing countries like Nigeria. It is our inability to maintain an independent exploration and production of petroleum, even if only a small quota, that will be the problem if we are suddenly abandoned by multinational oil companies in the near future.

Fishing is regulated. Wood cutting for furniture is regulated. Coal mining and crude oil should be regulated. To think that energy mix involving crude oil is in the past is myopic. No matter the emergency, the practical truth is that, especially for developing oil nations, it may take many more years to be completely rid of crude oil. The many cars, oil field installations, new airplanes, will not just disappear. It's a question of regulation, production quota and abating of pollution, during the unavoidable advancement.

Imagine we find a way of burning fossil fuel that does not pollute? That changes the equation, yet it is not treated as an option by some environmentalists. The real argument should be on how. That is why we need more concerned persons in the industry at this stage. While the major oil companies are divesting, a new type of man-power requirement is necessary, to get into the industry and participate in the final stage of crude oil exploration and production. This team will be HSE conscious, applying best practices, able to integrate other forms of energy in their quest, and form the basis for a smooth transition to the new energy world.

The Catastrophic and Fatality Tendency

This is known as *"fuite en avant"* or "headlong rush", which may make a desperate and fearful soldier to 'escape' to the warfront. Some people are adopting the catastrophic tendency. Most of them have no plan or strategy for implementing energy type replacement. They feel the world has moved on, away from hydrocarbons or fossil fuels. They are jumping into the sea because the abandon-oil ship-alarm has sounded. A new set of environmentalists protest and disrupt oil exploration and production activities, including pipelines. Professionals in the industry have argued that such protesters are hypocritical when they use the resource and forbid others to do the same. Their advocate believe that if there was an alternative, then they would have a choice. How quickly can we have such complete replacement for fossil fuels?

Reacting to the Article on the American Association of Petroleum Geologist (AAPG) News, by Heather Saucier, Explorer Correspondent, in December 2017, titled *Professional Protesters Threaten Energy Infrastructure*, Michael Nelson said that fossil fuels have elevated humans to an unprecedented standard of living around the world. The main problem the world faces is energy poverty. A satellite image of the world shows Africa and other energy-depraved regions of the world in darkness. There is a correlation between energy and development. People with little access to energy are impoverished. They need more access to the energy resources that developed countries take for granted.

This line of argument welcomes that those who work in oil and gas will continue to provide the energy needed to grow the world economy without harming the ecosystem. Chris Bischof said that in fact, the producers will produce, and the consumers will consume, until a better alternative works its way into the market. Government is incapable of creating a better alternative. There is no legislation that can make the Sun shine 24 hours a day or cause the winds to blow steadily at all hours. Which is why we are going to remain consumers of fossil fuels until a few new Einsteins find a new way to power the world.

Watch the Masters

Main oil and gas players are coming to terms with the idea of an oil peak, where the world reaches a maximum consumption after which there will be a plateau and eventual decline. The oil industry giant, BP, in its recent annual *Energy Outlook*, suggests that self-driving plug-in electric vehicles and travel sharing will reduce oil demands by 2040, according to its most recent annual *Energy Outlook* report. The forecast foresees that global consumption will plateau at around 110 million barrels a day. It does not say how long that plateau will last. This does not necessarily mean an end to oil and gas production, especially as there have been similar forecasts in the past decades suggesting that oil reserves should be depleted by now.

Some IOCs and the Western oil technology countries are leading the drive for new renewable energy. The difference is that they are using resources, heritage or finances generated from oil technology money. Norway is leading in steps to non-fossil based cars, but they have also saved and are still putting money from crude aside for their children's future and enough to build the new energy sources. This energy future seeks additional alternatives to meet the ever-growing energy demand. What is Nigeria doing as a country?

Those who lead in the oil sector do things differently. They act where no one sees; they do what no one is doing. Total Upstream in Nigeria has just commissioned one of the world's largest FPSOs, for production storage and offloading in the Egina field. The FPSO is the largest ever to be installed in Nigeria and the largest by Total Group worldwide. From the indigenous perspective, the Dangote Group is building a new refinery and petrochemical plant in Nigeria, which is a major investment for now and in the future. Their cement plant uses coal. Dangote obviously thinks differently and will get different result. These oil installations need manning in our own country for many years to come. It is one thing to talk, another thing to be ready and act.

Why do IOCs still apply stringent confidentiality to their technical information? These data are still unavailable to students, even graduates. That means there is still value in oil and gas exploration and production. Those who care do less of the talking; they are working in the background. The overly optimists run to the media and propose a change without a plan.

In a publication titled *The Dark Side of America's Rise to Oil Superpower*, included in the January 29, 2018 issue of *Bloomberg Business Week*, with cover: *Good For The Us, Bad For The World*, Javier Blas said, "it sounds good, but be careful what you are wishing." That article emphasised that by the end of the year, the US may well be the world's biggest oil producer. With that, America takes a big step towards energy independence. That oil is mainly no other than shale oil, the very one some fingers even in the US are pointing at, as not treating the planet so kindly. Yet, their interest seems to supersede and override that concern.

For the US, we are talking of eight-digits of oil barrels, more than 10 million barrels per day. Nigeria is doing 2 million barrels per day. We have oil shale in comparative lesser quantity and much more of conventional gas, which is relatively a cleaner type of fossil fuel. We can refine the minimum we need to power our own industries, generate electricity and stop importation of petroleum products. Independence in this vein alone can employ thousands of skilled people and reduce the dollar demand.

New career opportunities exist in the petroleum industry, as pioneers of the new energy transition and to ensure clean operation of oil fields. We need new recruits on the side of the nation, whose careers will involve a closer control and ensure that good practices are applied for safety and environmental protection. That is why we have to reconsider what we sign. As emphasised in that article in the *Bloomberg Business Week*, shale oil isn't any longer just about grit, sweat and luck. Technology is key. Geologists use smartphones to direct drilling, and companies are putting in deeper and longer wells. At current prices, drillers can walk and chew gum at the same time – lifting production and profits simultaneously. People saw it as primitive, in the past, but fossil fuel is also gaining a new technology. The same technology driving oil and gas is driving new energy technology.

Demystifying the Energy Chain

Energy is the ability to do work, by stored or moving force. It could come in the form of heat (thermal), electrical (charge), light (radiant), chemical (reactive), nuclear (radioactive) or gravitational (potential) energies. The sources are either nonrenewable or renewable.

Nonrenewable energy is energy that finishes with use, such as fossil fuels (gaseous, liquid or solid). Coal is an example of a solid fossil fuel. Nuclear is considered a nonrenewable energy, because the radioactive minerals can be depleted.

Renewable energy does not finish with use. Examples are hydropower, wind, solar and geothermal. Geothermal energy comes from heat stored in the Earth's subsurface. A particular energy is termed as renewable only because of its abundance in nature, such as those that occur as biomass in wood and plants, and those produced by man such as ethanol and biodiesel. All organic wastes, including sewage, are also biomass. However, rightly considered, biomass and plants are not renewable when we do not replant the amount used, except to the degree that coal and oil is renewed by decay of plants over millions of years. The quest to move away from fossil fuel hinders this insight into the similarity between fossil fuel and biomass. The other difference is the condition and time of renewability, which is difficult to meet and higher for petroleum products. It is also argued that the carbon footprint, or total carbon dioxide or equivalent emissions caused by biomass, is neutral. Carbon neutral means it only emits the same carbon content that it absorbed, not causing any overall change.

Renewable resources are preferred due to its less polluting effects, as most of it comes directly from nature already in the form of energy; all we need is to tap it. However, renewable energy sources are harder to optimise, meaning it takes as much energy itself to concentrate them. Either we reduce our energy demand or accept yet another costly means of renewable energy, or supplement with non-renewables. In which case carbon emission reduction will be the focus for fossil fuel. Gas generates less carbon dioxide than coal, while liquid oil is somewhere in between. In Nigeria, the use of gas, which we have in abundance compared to oil, is a way of reducing our contribution to carbon emissions.

The most important areas of world energy use are in power generation, transport and household heating. Today, coal supplies are close to half of the world's energy demand, but it is hoped that this will reduce to about a quarter of the energy supply in 2035, as reported by the International Energy Agency (IEA). The reduction from 40% to 15% does not mean reduction in the absolute amount, as the production of the other energy types are expected to increase. What is 40% share today may become 15% share of energy contribution tomorrow, if general energy production has multiplied.

The Simplicity of Electricity

If the principle used by Michael Faraday to generate electricity dates back to the early 19th century, then it should be considered as simple today. We have electrical coils everywhere in Nigeria, where we use small mobile generators. Our artisanal technicians recoil the assembly when it is burnt in our generator and electric fans. This coil or stator is made to rotate within a magnetic field to generate energy, with wires used to transmit the electricity away from the system. In a fan, it is the opposite; electricity is applied to produce rotation. In the generation of electric power, often what is needed is to rotate a coil using flowing water (hydropower), moving a piston by spark plug; you may use steam from boiling water when burning coal, burning oil or gas, or generating nuclear heat. That is why water is important in almost all of the cases. The equipment connected to turn the rotor attached to the coil is known as a turbine.

Any of the energy sources mentioned above can rotate a turbine, to produce electricity. If we could not build our turbines then, then we could have bought and found ways to rotate them. This is all that was needed to supply electricity to Nigeria since the beginning of this century. Wind turbines provide the same rotation. Geothermal heat or steam from the subsurface, which we do not have in abundance in Nigeria, can also rotate the turbine. The use of battery and solar power provides the same electricity by chemical means. We need to focus now on how to produce our own solar plates and wind turbines, to complement our power system.

Hydrogen, an alternative fuel type similar to battery power, is produced by steaming fossil fuel; this makes petroleum a major driver of hydrogen fuel technology. Hydrogen fuel is considered a zero-emission fuel when burned with oxygen. It can be used as electro-chemical cells or as fuel for internal combustion engines that power vehicles and electronic gadgets. As a petroleum nation, what are we waiting for to lead in this technology? As children, we built rudimentary wind turbines. Why can't we do the same as adults? Why do our people only pursue a career driven by salary? Yet only those who pursue a career with some dream or purpose ever get to the top in history. In our effort to inculcate creativity and innovation in young Nigerians, we make iESog trainees and Industrial Training students to do something with their hands, participating in the construction of lab-sized oil and gas separators, refineries, and solar and wind turbines, in our training centres.

The time has come for individuals to generate their own energy and supply to the local grid. We need to start building our own power-mix grid locally in our communities. The inverter, generator and wind, as well as solar option should be used by all neighborhoods. Our government has to authorise a decentralised power grid option, where persons able to generate excess can supply to the common grid. Having private local refineries for a community power supply, without needing to transmit the electricity over a long distance, would not hinder national refineries. Let's be open to smart options, smart grids and smart attitudes towards derivative processing.

Most of Nigeria's problem will be solved if we allow a regional or local equivalent of the federal structure in our economy and security. Local grids do not exclude central or federal grids; multiple independent power holding companies within a state will not prevent a national transmission agency from existing. But then the national options will have to compete. Government holding onto the assets, providing ambiguous subsidy schemes and exerting too much control seems to stop capable investors from coming in. Let the people have options, so that positive competition will drive a meritorious economy. This is a must if the new persons joining the labour market can have a fair playing ground and contribute their quota to national building.

Changing Education

The new education in oil and gas careers must be comprehensive, teaching not only the technology, but also heritage, as current and past history. Our pupils must understand where we got it wrong, the intricacy of oil and gas economics and how it affects international politics. These politics will affect renewable energy and we will lose again if we don't master it. We have to know how much we need to keep our country clean. At the final stages of oil production, it is important to leave the field safe. The locals must learn to be involved and do it properly; to be sure the transition to cleaner energy is smooth. We have to do this fast and now.

Things are changing – no time for extended formality, studying semantics, adding scale overly to diagrams, being pesky on all the rules. Advanced petroleum geosciences students need to focus on the crux. There is the need to lose the long unnecessary calculations and formulas. Over the years, the field of geosciences has been presented as if it lacked opportunities. This is why I introduce derivative geosciences to show candidates that there is a lot they can do with the products and by-products of earth resources. They must get their hands dirty, working on the raw materials to the crafts and designs that come from them. It is time to go straight from the skills, through the construction of equipment, to the full chain of exploration, production and refining of solid and liquid resources beneath the surface of the earth. No problem using the aids of artificial intelligence; let the robots help, but geoscientists or anyone at the base of the resource generation should follow the process to the end.

Going from the Infrastructure to the People

While we emphasise new energy, not much is said about what becomes of the current oil and gas workforce in millions, in Nigeria and worldwide. If truly we limit petroleum as a major source of energy, then it is time to lead the workforce gradually out of the sector in the long term. Some will transit to renewable energy, others will learn how to move on in life and succeed in other professions, while the others will remain in the industry.

Diversity and adaptation of personnel in an evolving industry are now necessary. That is what we should prepare for today as candidates training for the petroleum industry. You cannot be a one-handed geologist; you need to always consider what you have ...*on the other hand.* You need to have a geologist's, geophysicist's and reservoir engineer's skills. You now have to be rounded to work effectively and timely. As training is cut back in the industry, you need to seek skills acquisition solutions on your own.

Like many oil developing host nations, Nigeria just started to produce and refine its own oil. We have just stepped on board doing it alone with little external help. Then no more crude oil, it's banned, value is dropping, the planet is burning. That same tendency to stop a trade happened when we learnt to build furniture. Wood felling was banned. It is not necessary to argue if there is a conspiracy by developed nations to always set such rules. What is important is to be in charge and take each step circumspectly. You don't wait to learn when the game is over, else the game is over when you learn. If you wait to get the direction before you move, then someone is ahead of you. Human capacity building should start from day one.

By the time we learn how to build electric cars, the West would have moved on to something else. The problem is that we just jump, without taking our time to plan out our own long-term strategy. The "abandon and move on" sentiment, without proper homework, will always leave us behind. Good initiative demands that one asks: What can I start on my own? That is the spirit of leadership. Let us rather think energy mix, and use our current resources and heritage to get to that future. Let us develop a leadership attitude. Why not lead in regional biomass processing and have a centre that no one else has in the world? We have large biomass, in unquantifiable community and urban organic wastes. Why not lead now in biorefineries using these wastes to begin to replace conventional refineries? That means recycling the refuse, from every street in our neighborhoods, to energy in every state of Nigeria. If we wait, we will always wait.

8.3 Energy Mix and Global Warming

Many in our country were not aware that petroleum products could come from coal. Why didn't we explore that option when the rest of the world was mining coal? When Niger Delta oil was being mined, it was relatively easy to obtain, easy to ship, which was why we could not mine the gas in the Anambra Basin and elsewhere in Nigeria. Offshore oil can be produced and ferried away without any community complaining. We focused on the rent. We seemed to be seeing only one thing at a time. We drive one or two types of cars. We are choosy and once we select, every other thing is forfeited. The right thing, if we are moving on to newer energy, is to leave the older ones gradually so that we can have enough mix to industrialise. This is exactly what the developed nations are doing, just that they started early. They sound the alarm but put the date for abandonment ahead. The difference is what is done between now and then.

The energy mix will require that we benefit from the full by-product our resources offer at that moment. We should already have taken charge and utilised our oil, gas, refined petrol, petrochemicals and more. We can start from crude oil diversification – why did we miss the equation involving hybrid cars, biofuel, oil shale, coal-to-liquids? From where do we move on to diversifying energy types; why did we miss the nuclear equation and even wind energy? They have been there for years – over 50 years! Why are our youths on social media just saying what others are saying without giving it deeper thought? To properly transit and manage the expected new energy, we need to get the ones we already have right. It's a mindset, and a culture. All along, a good energy culture was necessary.

What the Popular Media are Saying

Below are comments I get from the social media. They are making some sense; I am only worried that the majority unfortunately are not doing the necessary homework:

- Why is Nigeria bent on petroleum resources? If we keep on using fossil fuel at the current rate, this planet will not be habitable in 50 to 100 years' time.

- Considering the fact that the petroleum industry is declining, we need to place our nation on the part to sustainable development in the face of automation and green energy technology.

- Investors are divesting from fossil fuel companies towards climate compatible development, even as bioeconomy is the next economic wave.

- In 2017, France and other countries set 2040 as the year petrol and diesel vehicles would be banned in the developing countries.

- Nigeria spend billions of Naira on crude oil exploration; even lost some officers during exploration in Lake Chad Basin. Are we in the 18th century?

- A decade from now, most cities around the world will ban vehicles that use petroleum products, ships are going greener and other oil-consuming industries are looking for alternative means to generate energy.

- The era of petroleum is fast declining and we should be more forward-looking and not get ourselves stuck on oil because it has done us no good. We are even fast running out of buyers. Let's please go back to agriculture.

- Petroleum is not the energy of the future. The nation should come up with alternative energy solutions. The oil and gas sector is the reason we are regressing as a country.

- Only one or two of our institutions offer full petroleum courses, which means our alarm to awaken the knowledge is late. It is late because many of those in developing oil host nations didn't know what they know today early enough.

- Africa is where it is, because we are always late. Before anything gets to Africa, it has almost been abandoned in the West. Most times we never put anything into action, we just theorise on everything.

- Oil is selling big but won't sell big forever, even though we have allegedly only discovered 30% of the world's oil reserves.

- Petroleum and coal are still very strong in terms of market influence, but other things matter too. We wouldn't want a situation where the world has moved and all we have is a bunch of "oil gurus".

- Global emphasis is on renewable energy due to climate change. In a few years to come, crude oil values will drop drastically to an unprecedented level.

- Within this transition period to renewable energy, conscious efforts must be made to optimise our rich conventional energy mix. More so, despite the ongoing clamour for sustainable energy, it's evident that oil and gas will still be in use for some decades to come.

- Developing countries with fossil fuel deposit should not jump the gun. What is much more encouraging is that natural gas remains one of the economically viable drivers of the hydrogen economy.

- Petroleum studies should be taught but as part of energy studies, in other aspects of renewable energy, to help future generation to not depend entirely on petroleum products.

- Nigeria must exploit her resources to make the best out of the oil and gas sector. Our fossil fuels, if sustainably managed, remain a blessing and not a curse, as is often regarded.

- The Western countries are unreliably dubious. We are beginning to take control. Whatever they can't control, they despise, and the fossil energy sector is one of these.

Join the debate:
Facebook.com/drlivinusnosike
Twitter.com/drlivinusnosike
Instagram.com/drlivinusnosike

The above list makes several points but those in the industry see more to it. Skilled professionals are needed during the declining phase of the oil field, and in the global decline stage. It is an art well understood in the petroleum industry. When we think of petroleum production, we can think in terms of power generation and internal usage, not just selling. We can think of the by-products used in petrochemicals, health and medicine, aviation, household appliances, and more, and consider that there could be new methods to reduce pollution considerably. The world over, those who are involved do not foresee a reduction in fossil fuel production amounts in the short term. What they propose is a reduction in the percentage of its contribution as more energy is produced. With increasing energy demand, there will be much more production of alternative energy, such that fossil fuel energy percent contribution reduces.

Imagine a new technique is found that renders petrol non-polluting. Consider the *catalytic converter,* a device used in exhaust systems such as the car exhaust pipe, to convert toxic gases (i.e. carbon monoxide) from combustion of petroleum fuel, to less hazardous gases (i.e. water vapour and carbon dioxide). It reduces the emission of harmful carbon gases to an acceptable standard. Will we continue fossil fuel production if it were possible to have it cleaner? Why is finding a safe way to use hydrocarbons no longer a priority? It seems, for the media, that everything is geared to throwing oil to the past.

What Industry Persons are Saying

The persons in the industry are not saying much. They are acting. I will, however, let you know what they are thinking. How they are planning the transition and why there is need for competent, skilled, free and fast-minded recruits in the petroleum industry today.

First, on the question of whether the Earth is about to burn, there are few unswerving geologists and earth scientists who do not think that 'global warming' is entirely man-made, but that there is a contribution from routine cycles that have occurred throughout the millions of geological ages. Our efforts to save Mother Earth will then only work if the external forces of nature are currently swinging on our own side. Evidently, oceanic *regression* and *transgression,* such that made the southern part of Nigeria (the whole Benue Trough) to be submerged millions of years ago, had nothing to do with mankind. When you point to the fact that the effect of global warming has skyrocketed in our days, they argue that it is a matter of scale where contribution doesn't make much difference.

This hardliners agree that man's activity is affecting the weather and even the climate, but that the scale of the impact is first local to regions of the Earth. So to them, controlling our activities is a necessity to abate pollution and aid good living conditions on Earth. Some external larger forces of nature lead to major geological catastrophes that are way beyond man's control. Emphasis is then first on the consequences of depleting irreplaceable resources; on not polluting the sea because it is harmful to sea life and mankind, rather than because sea pollution will destroy the ocean. As such, it's a matter of scale; impact on life as against impact on the Earth.

From the above perspective, our efforts to reduce global warming, based on short-term evidence of the rapid changes we observe, demands that societies will in their own regions abate pollution. In the long term, fossil fuel production should naturally reduce relatively with increasing newer energy mix, leading to it being lowered in the context of other pollution challenges. With good efforts on increasing renewables, we can put minimal exploration and production of petrol in the same context as other wastes that society attempts to manage.

With renewable energy such as wind and solar taking over, the residual oil and gas production becomes contributory and positive as an energy mix with less impact, yet necessary for oil-developing countries. Other waste such as e-wastes (from electronics) and nuclear waste, will be treated alongside crude oil waste, which good practices will reduce exploration and production. The reduction of fossil fuel production can be achieved by increasing production of renewable or other ingenious new energy ideas, not by forcing the rules.

Overall, the implication of this line of thought is that man is limited and more external forces shape the

Earth, over thousands or millions of years, than we can ever control. What we do today is important in terms of scale, but only in the short term. The world should be focused towards abating direct harm to life first, than save the Earth structure. This is not a popular assertion in either the public or scientific world. You may believe it or not, prove it or not. It is good you see the view from the other side, the view that is silent, the view that will be held by many when they join the career force in the petroleum industry.

No matter the noise, some people are involved and in charge of the world's energy needs. Even with the Western world leading in renewable energy, the people keep doing what they think is necessary in the petroleum industry to sustain the world's energy demand. Increasing energy demand makes it seem as if they have no choice; the very persons who only for ask alternatives sources of energy cannot wait. The industry insiders are not making any noise but acting.

Renewable Energy Wastes

In all this, I believe we should as well be aware of the oncoming responsibility as we go for renewable energy. I announce to Nigeria, to Africa and to the rest of the world, that in the near future, we will be dealing with major solar panel wastes and turbine wastes from wind power systems. That is in addition to the electric car parts needing waste treatment. Aeolian rotors, extenders, nacelles, towers, foundations and rotors, aerogenerators and solar gadgets, plates, inverters, and more will become large implantations as fossil fuel is downscaled. In Germany, 20,000 wind turbines have been decommissioned recently and 10,000 more are in line. Environmentalists protest as the residual of the components used for their fabrication is considered a major waste in addition to the aesthetic challenges. I hope these won't be shipped to developing countries as "fairly-used".

With time, more and more renewable energy wastes will flood the cities and their suburbs. This will use up domestic space, become environmentally hazardous, both in terms of waste and in terms of aesthetics in the future. Will there be a new solution, making the current renewables defunct? Back home in Nigeria, we may not lead in that quest and will again be held back. The hike now is on space exploration; a growing field for geologists, requiring geoscience experts to lead the rise on commercial space exploration. We may not all be astronauts, but how many in the social media follow what is going on inside, technologically, within the top industry? What if in a few years time a special resource from space replaces the current renewable energy, giving it an obsolete value by then? If you cannot beat them, at least maintain an energy mix while you join the campaign until we get there.

Continuous geological exploration on Earth, geochemistry and sample analyses, most of which have been advanced by petroleum studies, may be useful in the future with a resource that will come from outside our planet. The technology of unconventional hydrocarbons came from that of the conventional, just like sedimentary and structural geology, needed to mine coal, remained relevant to the petroleum industry. Similar electrochemical cell or battery technology used to start a car is now used to entirely power electric cars; similar technology used for gas turbines is now applied to wind energy. Similar principles of electromagnetism, used in microwave cookers and TV signals, are now applied to solar panels.

Get Involved and Moderate

As has been emphasized overly, let us then not suddenly and completely abandon the heritage of fossil fuel, not without due preparation and readiness for the next wave. One needs to climb each stage in-between, rather than import the new tech-gadgets we cannot manage. Check out the American Association of Petroleum Geologists (AAPG) publications and conferences on these subjects. They are talking about super basins, new resources and future directions in geosciences exploration. They are involved; they see where it is all heading. Ask the masters. Then add your own values, even if you do not fully agree with them.

When biofuel started becoming popular, I was worried and argued that we should not destroy plants for fuel at that scale; it amounts to the old utilisation of nature for mankind, which is not positive if overused. When biofuel worked in the West, few complained, even when it seemed not sustainable on a global industrial scale due to using up food or plant and organic matter reserves. I came to the conclusion that there are trends and waves in society, but what we propose today as an energy mix is not far different from antique windmills, use of the Sun to dry and cook in the ancient days and use of fossil oil slicks for lamps since the Stone Age. They did not harm the world until man's insatiable quest led to overuse and damage in excess of basic utility.

Biofuel, nuclear, fossil fuel, wind energy, hydrogen and electricity will profit man if used and exploited moderately, as had been in times past, before over industrialisation and the capitalistic craze. If the energy mix is adopted and no single resource is exploited to the extreme, then small-scale local exploration and production in the short term, as we transit to better energy, will not harm the planet. The benefits will outweigh the disadvantages in developing nations, who may be given more time to advance. A mix will considerably reduce the world's dependency on fossil fuel.

Assessment of the Climate

For me, there is a call for concern on the climate and a harmonious remedy we can all adopt by diversification of energy types. You can learn more on the assessment of the climate from Intergovernmental Panel on Climate Change (IPCC) publications. Among arguments for global warming are:

- There is a general temperature rise and oceans are warming.
- The ice caps are reducing and the glaciers are retreating.
- Arctic sea ice is melting and retreating and the water is flowing into the oceans, leading to sea level rise.
- Events of both cold and heat occur in their extremes and so are natural disasters related to the climate.

The consensus of the majority is that global warming is a fact. The questions are whether man has caused it or if it is a natural phenomenon, a normal cyclic change or the destined end of the world. I advocate going back to the way it was, and adaptation of new energy to abate pollution:

- Use petroleum in moderate quantities as part of man's energy mix.
- Use existing and new technologies to move from one energy form to another, from fossil fuel to renewable resources.
- Developing countries have to do their homework and find alternatives from within, not wait for the next wave of gadgets as solutions.
- Anybody aiming for a career in the petroleum industry has to know these challenges and be prepared to solve them.

Oil is one of the causes of atmospheric pollution; there are other causes such as cement production and deforestation. There are also the natural trends beyond the control of man. If you are a believer, then it may well be the apocalyptic end of the world or the biblical prophesies of the last days!

8.4 A New Kind of Workforce

If we had continued to mine coal and develop its by-products on a limited scale, we would not have only added to our economy but also would know how to protect ourselves from any harm while judiciously moving on to petrol in the last century. We were not ready and there was no workforce ready for the transition. It became an abandoned resource, while we did not master or independently produce and refine petrol. This also applies today.

If we do not prepare a new type of transition workforce, who require different coaching, there will not only be a lack of manpower in the energy industry in the short term, but also a forced abandonment. That is, a situation where the Nigerian oil and gas industry collapses, not because we want it to, but because it was abandoned. That is why IES Oil and Gas training www.iesog.com was founded, to use our oil and gas heritage to transit to renewable energy. Let it be our indigenous effort towards the same aim, which every other main player is attempting to achieve. All the supermajor oil companies are contributing to the transition. A recent Bloomberg report discovered that Exxon Mobil is investing in hundreds of projects in renewable energy, leading in the microalgae projects and conversion of waste lignocelluloses to biodiesel. The other ventures are not yet known to the public. Back home in our country, many are waiting to import the products when they are available.

While working in the Total Group in France years ago, I was amazed to drive past some of their research sites and assembly plants on renewable energy, on the outskirts of major cities. Recently, Total S.A. formed a partnership with EREN, now Total Eren, for the development of renewable energies – mainly wind, solar and hydropower. It obtained a licence to use 450,000 tons of raw vegetable oil for feedstock supply in its La Mède biorefinery. The Total Group's slogan is "*Committed To Better Energy*".

Royal Dutch Shell is leading in biofuel, hydrogen fuelling stations and actively working towards a low carbon energy future, promising to cut down its carbon footprint by close to one half and down by 20% in about 15 years time. They are carrying out other research, the knowledge of which is not yet available to the public. BP, Chevron and ConocoPhillips have also invested in renewable energy at various scales. The oil and gas industry is expanding into renewable energy, power and gas and so are the people working therein. Do not worry, there is still much work to do. The duty to transit to renewable energy is that of the petroleum industry; this is the message I want you to internalise as you start a career in the industry!

Concerns during this Transition

There are genuine concerns on both sides of the debate on the role of fossil fuel during the transition to renewable energy. Some arguments suggests that we should put climatic changes in context of geologic history records of extreme global warming and cooling cycles, beyond what man can impact. However, the current global warming campaign is not on the same scale as geological events, but a call to avert sudden rapid changes that may be man-made. It all comes down to what we are doing, alongside the campaign, to not only survive but also to lead and benefit from the transition. Some comments on the social media, arguing from an outlook on a better planned and self driven transition, are:

- Most times, we in Nigeria talk the talks, while others are busy doing the action we import. Every country leverages on what it has and builds on it to go further. It should be a block-by-block building approach; attempting to jump is why we fail.

- It is only now that developing petroleum nations have started to clamour for independence and local content control, that it has dawned on the world that petroleum is doing us harm. Nigeria has to look inward to look ahead.

- Petroleum is what you have for now; we need to learn before embarking on energy transition. Nobody says petroleum is for the long-term future, but if we are still importing petrol in a country like Nigeria, there is a problem.

- You should always be a forward thinker, to understand the dynamics of the world energy tussle. If we do not understand our current resources, we will end up importing solar panels, hydrogen and biofuel until it is again discarded.

- The worst is that we don't get it, even when it's gone. The reason we fail over and over again. Let Africa know her history, building from it in everything and in every way, not chasing the wind ahead.

- Unlike agriculture studied at all levels of our education, we did not adapt our education to our resources. That's why for 60 years we hardly mastered independent production of petroleum.

- There should have been national oil centres, like national space centres in the Western countries. Our pupils should have visited the national oil training centre, so they will grow to know what had become the mainstay of the economy.

- Although we are in a final stage for fossil fuel, that could take anywhere from 30 to 70 years! Even if only contributing as a little of the energy mix.

- While the debate is on, the US currently adapted its energy by developing oil shale. Nigeria has just developed the Liquefied Natural Gas (LNG) that harnesses our gas. It is not over yet and we must be real.

- Elaborate home-based petroleum economics studies, taking cognisance of international trade, will expose why oil and gas industry has not developed host nations in some parts of the world. This then will help us not to make the same mistakes in the drive for new sustainable energy.

- We need to see what we miss by allowing time to pass blindly while our resources last. Petroleum does not exclude green energy, on the contrary we need it to move on to newer energy. Those countries that lead in automation and green energy are doing just that.

- Yes, we need business education, industrial, manufacturing, practical production and engineering education. But how do you power that? We did use our petroleum energy to power our industries, as was done by industrialised nations who are now moving on to something better.

- There are people who dedicated their years of study in petroleum engineering and related courses, who didn't get a niche in the industry back home because, until now, we do not have an independent capacity to harness our full resources. Only selected persons share a place with foreign expatriates.

- Until recently, about 90% of the jobs are taken by foreigners. You see only about 60% of them in Nigeria, but about 30% more work is domiciled abroad indirectly. Reason being that we are not in charge, we lack the capacity, or so it seems.

- Our institutions only get the flavour of the metier. The few who study abroad for a master's degree are trained to do one thing. As the world diversifies, there is a vacuum left for local skilled geoscientists and engineers to occupy during this transition.

Again, you may join the debate:
Facebook.com/drlivinusnosike

Twitter.com/drlivinusnosike
Instagram.com/drlivinusnosike

The new kind of workforce, needed in the petroleum industry, will work within the available time to sustain our energy and transit us into the new world. I am showing you how to join that workforce, how to get in and get out at the right time. Your role of helping to transit yourself and your country into an independent energy future becomes clear as you understand the strategy. It is time to join the team and discover the new human and energy potential hidden in conventional, unconventional and new energy resources in Nigeria.

New refineries are being built in Nigeria for the first time after many years, including major and modular refineries being pushed for by concerned local elites. Major oil companies are selling blocks to local companies. Should all the oil fields be relegated to Nigerian players, who will do the jobs? Will you not be involved in putting your house in order when the visitors are leaving? Will you leave your farmland fallow because the farm lords are leaving? Like never before, we need to learn critical things about a career in the petroleum industry.

While the language is changing to emphasise energy, we need to remain focused. You already know more than the average employee, even in the industry, who has not read this book. Don't leave our craft and chase the wind while others fly in it with their craft. The persons in the media, the world, are right about renewable energy and new energy studies. We all promote renewable energy, including the oil giants, but not to a sudden exclusion of current energy source. We are doing something on the ground and we know the challenges, which is why we advocate you to prepare for the exit. Your career in the industry today involves playing a role in adequate preparation.

Many back home join the call for renewable energy without understanding how it works against them if they do not do their homework. In Nigeria, petroleum is perhaps the only resource that would have given us what we need to move on to the new energy future. We need to optimise all facets of resources, for the energy mix, knowing it is a transition that has to be smooth. Let us take control and have the power of entry and that of exit from all the developments that drive our life as individuals and as a nation.

Those who want to join that special workforce will scoop up the last oil and do the cleaning up in the oil and gas industry. They will employ derivative processing and derivative geosciences. I encourage the geosciences community to expand the meaning of their profession, from mining to the processing and manufacturing of the end-products of mineral resources. It is a call to full participation in the food and energy chain. They will have opportunities where others do not. They will see what others failed to see. Most importantly, they will have a different mindset, not that of settling, but coming in to conquer and move on.

I have tried to show you how to get in, make a point at this critical moment and get out of the petroleum industry or become an energy giant if you so wish. Here is a roadmap to freedom; an exit strategy with opportunities for individuals and for nations.

Chapter Nine

EXIT STRATEGY

There is a trap in the oil system whose only escape is to have an exit strategy at the time of entry.

9.1 Technology Transfer is Timing

Within the next few years, Nigeria will be able to produce all kinds of petroleum-based cars, but by then, it will be too late. Cars using other forms of energy have already been produced throughout Europe. It is not about gaining the technology, but when you get it. If the big oil corporations transfer technology in such a way that takes us 50 years to learn, then it is too late. There will already be something new to pursue. We then jump to that next one, a rush hour approach, which is the same as that applied now due to oil resources running out on the one hand and losing relevance on the other.

Understanding the timing matters. We need to have a sense of synthesis and adopt a comprehensive approach to problem solving. As oil prices fluctuate, the energy multinationals consolidate on solar and wind energy, which they have been working on gradually over the years. Our national authorities seemed to have remained at bay; it is only now that they are making efforts to ensure Nigerians are dominantly involved in the upstream petroleum upstream sector. It might be too late to salvage this aspect of the oil and gas struggle of which many Nigerians are not aware. Time is running out.

One secret of any enterprise or contractual undertaking is to foresee an exit right from the entrance. It is always better to have the option of stepping out. We depend on petroleum as if we cannot survive without it. It is necessary to have an exit strategy, both as a people and as individuals. Employment contracts and collective trade agreements contain clauses on voluntary end of service, termination or early retirement. It is the lifestyle of oil workers that makes it impossible for them to quit, however much they may wish to do so, except when they switch jobs. The dependent nation they come from enslaves them to the control of other sovereign states. The processes of breaking free for individuals and nations are similar. What can we do as individuals and as a nation to break free of the sole dependence on petroleum industry?

We can strive for freedom, technologically and financially, and solve the energy problem first. Sooner or later, individual oil personnel will need to think of the great exit from the traditional oil and gas campus-style work setting and associated intrigues. With changes in the industry and in the outer society, where everyone is talking of a greener side of life, young and old oil workers find it more difficult to badge-in and face their workstations all day. Yet many of them lack the prerequisite strategy to freedom and the opportunity to participate in the energy transition.

9.2 Synthesis for the Subsidiaries

I remember that as we left for our training in Europe many years ago, a dream was sold to us that on our return there would be an equivalent technical synthesis centres back home, to the technical hubs in Europe, with a mix of R&D and operational base, where people of broad mind could synergise to advance the industry. Many years after, it turned out that this dream was hard to come by. Everyone was looking across the horizons beyond the sea.

I once spoke to a career adviser, who at first argued that Nigerians have been trained to become capable of running the petroleum industry. He showed me the training catalogue as proof that Nigerian staff have been undergoing different forms of training for the last 15 to 20 years – consuming one of the highest training budgets of the subsidiary. I pointed to the same catalogue and showed him how the training organisation was meant to divide personnel on specifics. They were learning one job at a time, only enough training to work on a given task. Such training came in short, punctual, narrow windows adapted to different staff profiles. There was no focussed apprenticeship, where a new local employee is trained continually by an experienced staff till perfection.

The career adviser accepted and explained how it was difficult to suggest the idea of having experts in Nigeria; that the tools, the collaboration and the will, were not there. We came to agree that it is not the absence of individual training, but a lack of a comprehensive vision or synthesis, that makes it difficult for indigenous staff to become truly independent and capable in the petroleum industry. Perhaps these training courses were not meant to set the employee free; the local learners did not get the whole picture early enough in their career, to trigger the motivation and passion to operate the whole sector.

On the national side, there was no sense of direction for aspiring young career persons, in a sector that has been the main source of the country's revenue. The national oil company was run by politicians who preferred politicking to professionalism. The smart and well trained Nigerians would rather join the IOCs. That is why there are no functional museums to showcase our oil and gas history or heritage since Oloibiri, Nigeria's first commercial oil discovery. There are no theme parks or any form of emblem that brings the masses to this consciousness. The upstream petroleum sector is big, one that has been central to our economy. Yet it is run in the shadows, such that many Nigerians only see rig structures on TV or in the newspapers; some of these are offshore – as distant to them, as another world in outer space.

In a system already in motion, inertia makes the young employee to flow with others and contend with working for their pay. If only they consider their share in the sufferings of the poor masses, partly due to the corruption resulting from the rent-based financial allocations in Nigeria's petroleum sector. Under pressure, the majority struggle for survival, many serving the purpose of a system they barely understand. Employees do not see the wrongdoing of inaction. If the injustice doesn't bother you, you will not go that extra mile. Everyone wants good, but few hate evil.

9.3 Time for a Mop-up Exercise

Is it worth still paying attention to oil and gas while the world searches for a cleaner, sustainable alternative energy? The answer to this recurring question is yes, because the process of transition will last for between 30 and 60 years in different parts of the world, if we consider residual forms of fossil fuel use. Rapid technological advances can change that time, to be shorter or longer if they abate pollution. For an oil field, a reserve management strategy is used to optimise remaining hydrocarbon as well as properly return the field to its *as-it-were* state. Then comes a post-mortem to learn from the good steps as well as the mistakes made while developing and producing the field. Nigeria, and indeed the world, needs to outline a strategy for a declining energy resource and plan to optimise remaining oil and gas reserves before abandonment. Not doing this, by the people whose responsibility it is, is a wrongdoing by inaction.

Usually for an oil field that is experiencing decline, the objectives will be to:

1. Completely sweep the oil
2. Optimise technology
3. Diversify production approach
4. Look for alternatives
5. Leave in clean good condition

This is what I call a *mop-up exercise* in the larger context, of what the world needs and what defines the role of oil and gas recruitment candidates today. They will have a different mindset from the old oil workers who still idealise a comfort zone. They will pay more attention to environmental consequences and use the best HSE good practices that are in place to:

1. Do a final exploration and production chase
2. Begin to think modular solutions
3. Achieve diversification of resources in Nigeria
4. Aspire for freedom and look ahead
5. Research and develop clean sustainable energy

At this stage, we will be careful to abate pollution and go that extra mile to ensure others do the same. An effective foreign investor's check is impossible without involvement of a knowledgeable local hand in oil provinces. We need an investor-friendly environment. Let's not pull down a system if we cannot rebuild it. Leaving our responsibilities to others is not the solution either. Everyone has to be carried along; let us all contribute in the good will and by our actions. A lot revolves around the financial capacity of a nation and of the individual. While the golden rule reminds us to treat our neighbours as ourselves, it turns out, as they say, that he that has the gold makes the rules.

The precepts of success are found in the written and unwritten creeds of insider investing, of wealth creation and of economic development. The oil industry workers felt secure and uninterested, but times have changed. As you pursue your career in the petroleum industry, I invite you to discover the principles of freedom outlined below early enough. You too can invest and live a free life; to do what you want to do, not to entirely pursue a career under constraint. Remember, every exit leads to a new beginning. Let us exit from the past, let our future begin today.

9.4 Financial Freedom

All through, I have laid emphasis on the fact that human survival is at the heart of men's allegiance to systems. All the methods of control and oppression hinges on this one need of mankind. Freedom starts from individual and national ability to make ends meet without collusion. That is why it is important to grasp the basics of income, expenditure, savings and investments, even for the oil worker. We will not do ourselves any good if we keep talking about all that needs be done without empowering the persons who will implement the change.

You cannot make your desired change, or exit the system you are in, if you cannot sustain yourself financially and materially. This is why knowledge of wealth creation is important for the poor oil host country and the individual oil worker. Until we deal with the money factor in our lives, it is difficult to become available to add any value to society. The money factor is a bane to not only those who lack money, but also to those who are controlled by money. Oil and gas workers get a lot of money, but it sometimes steals away their freedom.

I struggled between my day-to-day job while I worked towards the freedom to live what I believe. How I achieved that is what I share with you as the process of financial freedom. Faith is necessary; you need to believe that all your needs will be taken care of by divine providence while you do the needful. It's the individual citizen that makes national changes, but the oil elites cannot contribute a change if they are burdened by an inability to survive a day outside the company system in which they work. They must have enough ability to survive before they can have a say, because it is difficult to tell someone to see a fault in the system that feeds him. For the new recruits, the plan to exit must start right from when you first get into the oil and gas industry. It holds true for both the nation and the individual.

First is to know why you need wealth; for some people, it is to get significance personally, for others it is to do good to the society. Something has to be very important for you to achieve it. That becomes your driving force. We are concerned about freedom as a nation, as individuals operating in the petroleum industry, to be free to do what we want to do, what we believe in; to not live our old age in regret. The principles below hold true worldwide for financial breakthrough. To make and maintain wealth, you need to plan and work strategically towards it, over time. In oil and gas, it turns out that many are missing the idea that money alone is not enough to build wealth.

Let Your Goal Provide the Means

One error people commonly make is to wait to have the means before they decide on a goal. Oil company personnel keep working and hoping for that day when they will have enough resources to embark on their dreams. They want to have money or a good job before they dream big. The more they wait, the more the changes, and so the changes of their dreams. However, it is to the contrary. Dream big and your dreams will show you the means.

Opportunities pass by you daily, but you cannot see them because you don't need them at that time. You think you do, but not backing your vision with actions shows you don't. You must think and look back at your original passion or goal in life. You need to find out that one thing, that the desire to achieve in your lifetime generates all the energy in you. It must be so important to you that the enthusiasm will keep you going once you are on that path. You had spent all your energy doing this one thing, when you had the time.

If for example your dream is to build a stadium, go and choose a place to build it. Find out the cost. Obtain and study the documents you need to process before you build. Do a sketch of the stadium. Look at it daily and think of how to take the next step. When all your desire is in that direction, you will see situations arise to show you the next step. Act on those opportunities and you will be one step closer each time. Waiting to have the money before you start is why you never will.

Spend Less Than You Earn

This is general advice to everyone aiming to be free from subdued labour. Our country is borrowing and spending on recurrent expenditures. For the oil personnel, no matter how hard you work, if you squander all you earn, or borrow to spend recurrently, you will not be prepared for the future. It is a mistake to change your living quarters and change your car and even change your spouse because you got a job in the oil industry. Those changes will continue with your changing income. You will remain enslaved by your job, even if you do not like it. On the other hand, those who pursue their dream may maintain their work with joy because they appreciate what they do, not because they are solely constrained by the need for a salary.

While this injunction on spending looks simply true, it is easier said than done. You could tell how much people spend by knowing how much they earn. In the oil and gas upstream sector, despite huge monthly pay and annual benefits, the majority live above their income by taking out loans. The expectation of the job, lifestyle and proximity to high rise urban offices often result in over expenditure, which ever keeps growing as family and kids come along. One way to ensure you are not trapped is to delay your migration from the low expenditure life to the high, thereby maintaining a low expenditure with a high income. It will of course require much discipline, and collaboration with your spouse, no matter when you start. Once you spend less than you earn, put the balance where it will generate more money for the future. This could be in an investment, fixed asset or special secure high interest savings account.

Spending less than you earn does not mean living meagrely. Most people who overspend in fact live poorly; they put their money into big flashy cars and houses and all those things that other people can see and then, they are broke, often borrowing to feed each month. When you spend only on your real needs, you have enough for the important things in life. When you do buy flashy cars, let it be because you are spending a little profit from your passively earned large income, not adding loans to your active employee-based salary or sweat from hard labour. Don't buy comfort with duress.

Earn More Than You Spend

If you cannot reduce your spending, then peg your expenditure threshold and look for how to earn more. This has become more difficult for our country to do, because we have limited growth in earning. For you as an individual, you need to save and invest any salary increase while maintaining current spending limits. Create more income streams, active (where you work physically) and eventually passive (your investment) incomes. Once you earn more, then put the balance into a basic (easy to understand) venture in your control, or that you create, which will yield profits in small but regular streams. Don't start something too sophisticated at the beginning.

If you do not like your current work, do what you want even while working; use the spare time outside work to pursue your dream. A day is made up of 24 hours; split it into three: one-third for your job, one-third for your passion and the other third for sleep. One-third of 24 hours is eight hours, which is the maximum most employers will officially ask for. Your passion should drive you to what you like doing most – adapt that to give service to others and to earn additional income.

Oil workers are particularly tied to their job, behind computers or on the rig floor all day. They resort to giving over their money to someone to manage, in which case the person or agency makes their own profit first. This is why it is harder for persons in the industry or other constraining work to break free. The solution is that they make more sacrifices by not spending the larger remuneration they earn as compensation for this demanding job. Ideally, the more you are paid, the quicker you can save to invest and retire early from your job. Unfortunately for the oil worker, the more he earns, the more he has adapted to a life of comfort and the need for more money. Oil personnel put their hands in so many material acquisitions and projects, thus making it even harder to quit their job and full salary, not because they love the job; some even after retirement.

Delayed Gratification

Delayed gratification is a major principle that every successful person has applied. Why many cannot spend less than they earn is because they cannot wait to enjoy the proceeds. Just waiting a little for your results to mature makes all the difference. As a general rule, do not spend your gain, but the gains from your gain. Imagine a farmer that cannot wait from the planting season to the harvesting time. He will dig out his seeds and eat them one after the other.

People behave as if they do not believe the future is real; in any case they do not feel it will ever happen. They either do not have faith or they cannot risk earning a wealth they may not be able to spend. Some think, What if I wait and use my earning to make big money and then die in an accident? They find it difficult to imagine that their wealth will go to their loved ones or shared by the government. So they would rather squander everything and ignore the future. Unfortunately, nature does not ignore, it waits for many who don't care by punishing them with poverty.

You need selfless generosity and faith to delay gratification. You may have to pay salaries for years to people you employ who don't make money for you, just biding time with them until you become available to manage the estates you struggled to build while you worked in the busy oil and gas setting. Just like the seeds, only those whose consumption was delayed eventually germinate to become plants and bear more fruits. Most professionals in the oil industry believe they work hard and deserve their pay, so they eat up all their seeds.

Reduce Leakages

As a busy oil and gas career person, you are often cut off from society and the few times you mix you are expected to show yourself off, demonstrating that you are rich. The consequence is that you solve problems with money essentially, which creates an ever-increasing dependency on you by persons in your sphere of influence, through all available avenues. Your resources leak away in the office, dependants' schools, builders and errand boys.

Trying to close up the leakages completely will not work, as people may work against you and get you even more frustrated. There were times I had to look away when I found my helpers, despite substantial benefits, were stealing my resources. What mattered was whether they were assisting in some ways as I pursued my dream. However, I introduced channels that minimised such leakages. I transferred money rather than sending cash. I used my spare time to visit construction sites. Many of my colleagues vowed not to build anything until they were on holiday, or at retirement, due to trust issues. Others gave over their money to costly construction companies that only build for rich people.

You need to know that once you are absent, you spend more. Track where you encounter most of the loss and close them up to a reasonable extent. Don't give up on your dreams; people will always attempt to cheat you.

If someone you paid no longer has the money to deliver a given task, then even if you jail him he may still not have it to pay back. Sometimes you don't have to probe further when it is clear that the person has regrets and is willing to make amends. Persons who are already embarrassed can make you want to have no more to do with them. Save them, ignore the minor loss and proceed as if everyone is willing to make amends.

However, when a defaulter is not contrite, then expose and fully dissociate with him or her. You should turn away if there are signs or a trend of betrayal. Go on to work with new people; try again. It always seems that without one person you are doomed. I find that every end of a journey opens up to another. You can maintain the benefits of having people around you, but remove the losses due to negligence and distance.

Spread Your Net

Oil industry professionals have a narrow stream of income, just their single salary; at most they have a share in some stocks or a mutual fund they have seen on TV. To have multiple streams of income, to increase your earnings, you need to have a multiple spread of investments. Diversification of investments helps you to cushion the effect of any of them going down. You never know what is going to work at first. Don't put your hard earned cash in one basket. Don't save all your money either, as inflation may wipe it out.

Another problem with oil workers trying to invest is that they attempt only the thing in vogue, the one that everyone approves of, the one that is fashionable. Often those are the over saturated deals ready to burst. How do you decide on where to put your money? The answer is, do not decide. Put it everywhere or at least in as many places that you believe in. When you put your money in things you understand or have created yourself, you are called an inside investor. It will take time, just like a farmer weeds and irrigates, then waits for his crops to develop until the harvest season. The quicker you use your spare time to supervise these investments, the better for you.

If you want to go further, then put your money in small rental properties, event centres, farms, shares and bonds; in commodities such as gold, in government treasury bills, in retirement plans, in different countries, or wherever you believe is okay for you. Start small and grow. Choose at least five different assets where there is less risk in the long term; build these assets yourself where possible. As you grow older and plan for retirement, you should no longer put money into those investments that seem like a gamble. Building your own investment is the sure way to gain experience, mature and have control.

Take Value to Where it Matters

Giving value is the only legal way to earn wealth. Even when you trade currency, you are providing it to someone who needs it at that time. Value is only value where and when it is needed. The question is, what value can you offer, who needs it and where? Take service to where it is most needed. Unfortunately, many in the oil company employment do not have that flexibility to take something from one place to the other outside the oil field and office work schedule. Independent consultants do to some extent; investors must have this flexibility.

For example, if a cup of *garri* is ₦10 in your village, then take it to Lagos and sell it for ₦50. The people there will be happy they got it at all. You are adding value both to yourself and to them. Sticking to your one standard and environment is one of the reasons people do not break free financially. To spend less than you earn, don't spend where you earn. If you make ₦100,000 a month in a chic quarters of Lagos, do not use it to buy food in the costliest market there. Rather you should buy it from a neighbourhood further away, or from a market in the nearby rural market. Where you earn your money, all is geared to take it back from you.

Where do you choose to build your rental property and your retirement house? What will it cost you in that urban city compared to the cost in a smaller town? Will the profits be the same if you were to build the property in the big city and your retirement home in a less costly town? Oil workers who ignore this basic reasoning and do not take the simple steps to act accordingly on a daily basis, are stricken with the shadows of lack, in these days of black gold sector decommissioning, when they need to join others in the strive for survival.

Fix Your Assets

While you do want to multiply your money, shift it gradually to physical assets such as buildings, land, or other fixed assets that are not easily moveable or which cannot be easily snatched from you. This gives you confidence and empowers you to act, knowing you have something on the ground. Liquid cash gives only temporary confidence and can be lost at any time. With inflation, money in savings may lose its value. Physical investments such as property may fall in capital price but, like gold, their physical value remains the same. When they do bring in rent, the rental income increases with increasing inflation.

Check your investments to ensure there is always a residual that is constant, and will remain so in all cases. If you open a school, do you own the building? If you invest in farming, do you own the land? If you run a business, do you own the premises? Should the agricultural produce be destroyed, or the business goes down, what remains? That is a backup for the long term.

While you spread your net, focus in one or two States, regions, or the town where you intend to spend your retirement and ensure that you have the basics of life there. Focus on one or two investment or businesses, around which others pivot. Oil workers before now felt secure; they thought they were better off than those who earned less than them. Today everyone is in the same boat, thinking about the future. Don't ever be too sure. Plan, knowing that things may go differently from what you planned. Always be prepared. Fixed assets serve as your backup.

Love People and Life

Remember the purpose of learning the principles of wealth creation, to be free enough to contribute to our society. The oil professionals before now did not do their duty because they did not have the financial freedom despite years of work; they remained entangled with the fear of losing their jobs or not being promoted like their peers. With financial knowledge comes the freedom to act and be more involved with society. You can now pursue your dreams. Most dreams are centred on people and life. If your dream is not, then it is probably myopic and may not harness the power of nature to draw wealth to you. Look at other people who did something great, they all had to be involved with people at some time. Your service to people, to make life better for the greater masses, is the most important commitment that will open the door of success to you.

Countries that centre their goals on services and care prosper. Big dreams go with big service to many, which in turn create a link where life providence flows your way. It is from there that the good life you aspire to will come. The great aspirations of life create the bigger picture harnessed by men on the pathway to greatness. People who only toil at their table and workstations, as is common in most oil and gas companies, often miss out on the big other side of life. Live, work with people and open up to society, not merely interacting with machines and computer screens. You will then find the pathway to the success you had always desired.

9.5 Know the Right Path

Taking a leap of faith as a nation or as an individual to break free from encumbrance leads to changes in your life. You may not know if the challenges signal doom or if they are the pains of labour for a rebirth. There is the need for reassurance that you are on the right course. I have outlined some checks below in the area of control, timing, long-term, income stream, pressures and assets, as well as passion, to guide you.

Have Control

Nothing can be more important in the exit strategy than control. The best control is obtained at the time of signing any contract, which is when to ensure you have an exit plan. If you are an employee, are you happy and being carried along or do you have this feeling that you are not in control? Sometimes as individuals, we have to moderate our expectation and consider our input versus output in any organisation. If however it is obvious you are not in synch with the world around you, chances are that you are getting further buried than coming out of the hole. Each time you have a suggestion, from your peers or from your company, always ask yourself if you are in control of your decisions and actions. Do they go alongside your bigger dream?

When you have control, you can always start again and adjust even if you made a mistake. It helps to start afresh sometimes, a zero-based thinking, meaning you behave as if nothing had happened before. You can start today to build a new life. Each decision point is a new slate to write your future on. When you have control, you know that the past no longer exists and that the future is your choice. Put yourself in the picture; tell the story of the past to fit the present and the desired future. Act now as you will want it to be, not letting others act in your stead.

To be in control, you need to take decisions before they become urgent. Act on things that do not pester you at the moment. Those important but less urgent needs are the ones to focus on; they prevent the future emergencies and give you the chance to pursue your desired end. Urgent things cost you more. Anytime you are not in control of your life, you are losing to someone who is. If in that nice petroleum company setting you feel out of control, then you probably are not on the right course.

Time Zone

Our morning and night occur at different times, depending on where we are on the surface of the earth. For example, in China and in Nigeria, night and day come at different times. If you are out of your time zone, you find you feel sleepy during the day and awake during the night. Do you have a feeling that you miss chances? You are always late for an opportunity or it comes as soon as you leave? Do you often get it wrong, or things go the other way for you? Perhaps you have left your time zone. This happens when people have changed their original dream due to pressure, such as peer, family and societal pressure, or trying to do like others. You enter the time zones of those people unknowingly and don't fit in.

Sometimes you need to find your base. When you travel from one time zone to another, we say you have a jet lag. You feel drowsy and downcast. Getting back to your rightful place may feel the same, after treading so long on the wrong path. Like leaving your regular monthly salary in a company you think is wrong for you, to where you earn less but are happy. You can opt for a less structured income based on faith and individual initiatives. You can change the zone that is at odds with you, until you find yours. Then wait for your morning to come. Follow your heart.

Anytime you land, check that you are in control. Imagine you come from a remote town where people are hunters and can shoot an arrow from a great distance to kill their prey. You are in a stadium and you have two options, of playing darts or playing football. People feel football is classy, but not you. Coming from that remote village, you are better at shooting arrows; most likely you will play darts very well. You are afraid that you will be mocked and become helpless if you lose. What will you do?

If you play football, you may be like a fish trying to climb a tree; you are not in your world. Playing darts may make you the best dart player in the world, even if remuneration is low at first, even if you are temporarily alone or lack constant resources to start with. Pursue your dream in the long run; adjust opportunities to your dream, even if your exact expectation is not yet available. If you compromise with yourself, if you do not live what you truly want from inside, then you start drifting away to the wrong time zone. In your own time zone, you derive joy in what you do.

The Long Term

How far can you see? Do you have this feeling of only predicting your immediate future? It could mean you are not on the right path. Being on the right path makes you see tomorrow, you are reassured and you have passion. If we know it will be okay tomorrow, we are able to overcome today's challenges. If your entire life is driven by short-term objectives in one oil company, and you are tossed from here and there as the oil flows, then you are being jolted like the strokes of a piston until you wear down. Thinking for the long term starts in your own head, residing there until you see the silver lining.

As a young wellsite geologist, I already had a PhD and knew more about geosciences than my counterparts, most of who had only a degree in geology and were good at just what they did at the rig. I started with catching samples in the first week of my going to the rig; that is collecting samples from the shale shakers and cleaning it up for analysis. This one demanded only very little experience or technical competence, but it was a way to be useful on the rig when it is one's first time. It was a difficult and perhaps sapping task for someone with my geosciences experience. I had a long-term vision, so I didn't mind. I tried hard to not present myself as a PhD or experienced geologist. That would have made those technical hands and wellsite geologists enter their shell and leave me to flounder. Or they may well have antagonised me and lead me astray.

In the petroleum industry, the journey in one's career is a long one. Later while I was working at base, as I changed departments or teams every two or three years, I was content to know that I was gaining experience across the board and preparing for my future dream of running a geosciences firm that would present a new perspective of the oil and gas industry in Nigeria. Each time I joined a new team and met my new mates, chances were that some of them had been in the same position for about ten years and would be more experienced with just that job specification. They were happy to know better; some were helpful, others were bossy, but it never really mattered.

My openness in accepting to take up the next challenge, and the continuous tests I was subjected to, moved me around. Before ten years had passed, I had worked in six departments, eleven teams in several different countries and for more than twenty jobs: exploration, synthesis, operations, projects, developments and assets. Each time I joined a new team, I often had to go back to the bottom, but overall, for me, it worked for my long-term goal. In the end, I was doing what should be a specialist job across borders. It required patience and could sometimes be challenging; being in control within, knowing my time zone, ignoring the present difficulties if it served the future, were sure signs of being on the right path.

Earn Passively

What if you are unable to work with your hands today or your company collapses? Do you have any other source of income? If you are among those seeking to stand, to remain relevant when the oil and gas boom is over, then you need to start to seek a passive income today, no matter how minimal. Passive income comes from invested money, not by your physical availability as in active employment. With trickles of income while you sleep, you grow richer gradually. It demands sacrifice to put that income there and go to sleep; that is the faith nature wants you to act upon. If you buy a rental property and sell it when it appreciates in price, then you make a quick gain, but you will wake up the next morning without anything trickling in. There are times when trickles are just what you need, better than killing the goose that lays the golden egg.

For an employee in an oil company, seeking to make quick gains in external investments is not the way to go. You will not have the time to always get such chances, to monitor the market and to sell high. You end up falling victim once in a while, sweeping away your earnings when you attempt to focus on capital gains from buying and selling huge assets. Different things work for different people at different times, but it might be better to hold on to anything that earns you money, receive the small additions and expand such assets. They provide basic financial support and hope in times of trial; they grow with you and could make you rich in the future.

Earning passively in the petroleum industry is only possible if you have not given in to excesses of societal demands and a showy lifestyle while young in your career. If you do, you have nothing left to invest. You may start with investments that take less time, like stocks, while you retain your current job, or work until you grow your earnings enough to buy investment properties, vehicles for hire, or whatever works for you. It turns out that having a second or third stream of income that does not require your physical efforts is key to breaking out if you do not like a routine job. Some persons prefer and function better in routine jobs. Many of us have to work, and indeed gaining employment and working is key and necessary for most young starters after graduation. You work hard until you get to that point where your money, if you save and invest it, works for you as well.

Successful persons who had jobs still work in some way today. You may be an employee or an employer; no side is easy – it is about being where you should be over time. Career growth should lead to independence or willful work in a company, not constrained by luck. In a good world, people should naturally transit from active to passive income, from active hands on the job to supervision and to management, up until retirement with a retirement account to take care of them when passive or old. We are all different, some working a day job as a preference and others starting up a breakthrough venture. Everyone will not be the next Mark Zuckerberg, but even then, we should have a strategy while working for when we cannot work. Having some passive income, even if in a good retirement account for regular employees, is a sign you are on the right track to freedom.

Freedom gives you more room to relate or work willfully with an organization or other individuals. You cannot have an interdependent relationship if you are not first independent. Independent people are able to relate with others more constructively. Even as a country, we fail because the other developed countries are seeking interdependent relationship with us, while we end up being dependent on them. That is because we lacked independence in the first place.

External Pressures are the Opposite

Oil industry employees are used to pressure; they end up doing only things demanded by pressure. Once there is pressure on you to do something, however, it probably is not in your personal goal's interest. For example, if there is the pressure to pay tax or to pay one fees or dues, then it is first for the seeker. That does not mean it is wrong to pay those pressure demands, but that you should not do only that. There are other things that are your own goals that nobody will remind you of. Do those things first, they are your future. As mentioned before, successful people learnt to do things that are important but not urgent. By the time things become urgent, they cost more and may not be worth treating as such.

Many oil and gas workers live a reactive life where they struggle to meet with upcoming targets, events, meetings, expenditures, hardly ever doing anything not pressing, for their own personal future and freedom. Decide today to make a list of what is of topmost importance to you. Take them on, one after the other, in order of priority, and do the top item from the list each time before spending on anything else. Resist pressures from all other angles and choose your own stakes. Nobody will pressure you to go and plant a seed. Yet planting when you do not see the harvest is the only way to have one in the future. As mentioned before, salient peer pressure is one of the major causes of lack of control, and excesses, in the oil and gas working environment. Make sure you prepare for your exit, for your future first. Take a look at your daily life, notice how being so much under external pressure is not a good sign that you are on the right path. Let the pressure come from within, eliminating all external pressure.

You are the Asset

We acquire all the hard skills in the oil and gas industry while forgetting the most important, the soft skills, the ones we need to survive with people and not with machines. Remember yourself, the inner you and allow it to be transformed with love and care, and be enriched in wisdom. Feed that inner you with knowledge and positive emotions, and let it be what others see. Let them see your passionate heart, not just your external labour. Skills and love make you a leader in people's hearts. This will make life easier and attract all the good things of life to you.

Many technical personnel do not learn about human relationships, until years later, for the few who get to managerial levels and by then it is too late, as the training is again directed to only management of in-house personnel. Acquiring soft skills means working on you to work better with others. Prepare yourself to succeed with people, just as you gained skills to work. To be successful and connect with society, especially during and after a hectic life in the petroleum industry, you need to display agreeable behavior. You need to think about people and share their love.

Investing in yourself is key, because only you remain when all else is lost. You are relevant because of what you stand for and can offer. Develop that core thing you stand for, work to get yourself ready to achieve your purpose in life. Develop verbal and entrepreneurial skills and a charitable character. Learn skills, whichever one is in synch with your life goals. Develop your person; take care of yourself to take care of others. Have in you what you expect from your spouse or in your customers, so you can attract the right persons.

Many oil workers attempting to invest finish spending on external pressures from taxes and levies, from clubing and partying, from all kinds of solicitation and expenditures, before remembering their own dreams. Didn't we say that good dreams are centred on people? Some use the term, "pay yourself first". That should mean 'put what you earn first in what you believe', not necessarily that you should be self-centred. Prepare yourself for what you want; develop to share. If it is charity, if it is investment, if it is religion, if it is education, do that first – the ones that resonate with you and your future – stemming from you. See yourself as a gift to the world and fulfil your responsibility. Prepare yourself to be a hub for others, as good relationship with people is a source of success and happiness.

Dream Drives Passion

We started this journey to true freedom by saying that your goal provides the means. You cannot continue the journey into a new positive perspective in the energy industry if you do not have a dream. You will not have the enthusiasm to pursue anything with full passion if it is not a life goal. Thinking great and acting great will be possible if you can find your passion in everything you do, doing only what you find passion in, by living in such a way that it fulfills your dream.

Should you want to be a photographer, then learn to see your working screen as pictures. Let your mouse and keyboard remind you of how to edit photos. Whatever you do, have a dream. Be consistent and have focus. If it makes you work hard, then money should follow, rather than money driving what you decide to do. Most people fail because it was money that decided their efforts; they do routine jobs for wages. Great men pursue their dream to give people some special self-born service and are so good at it that people pay them.

When you do not do your passion or work your dream, you do things without attachment. You may be in a situation of rot, of brain drain, like a routine task just for money. It is not an easy step to take, when we attempt to make a change, because over time we create a routine that alternates from problem to money and back to problem. If you run to those around you in the same job for advice, you find people running their own course for help. Doing your dream as time passes is the first way to be free, to live your future and do what you have always wanted to do. Ensure you enjoy the journey as well as the destination. If you have a block industry and that is what you like doing, then plant flowers around it. Paint it in gold. Drink juicy icy cold drinks in a bath around there. Enjoy it, even before you make one kobo. It is about the process as well; don't wait for tomorrow, for today is the tomorrow of yesterday. Enjoying the journey is one way to surely enjoy the destination.

Someone described it as "paying as you go". Do those things you can now, do not wait. It might as well be to do what God had imbued in you, that responsibility that is only yours on Earth. The major reason I preach freedom is to free oil workers from the bondage of excessive dependence on the system. The entanglements of this world, as they make financial pursuits, increase with increasing work demands and lead some employees to eventually give away their souls. I preach freedom so that you can be free to live this life given to you by the Creator. Feel the breeze of freedom again, like when you were a child. To live it the way you will want to, within your life's purpose. Seize the opportunity. Call us if you need help or reach out to Elvee Centres for Personal and Professional Development. You are one step away, exiting your undue ties and enjoying freedom.

9.6 Give Back to Society

If you love people and life, and the wealth of nature has come your way, it was so all the while for you to remember those very people you left behind and make life worth living for them too. Find a way to make an impact on others as time goes on. Don't wait too long, as you do not hold the key to your life on earth. As they say, prepare to die empty. Give out all the talents and virtue in you, play your unique role.

You may be afraid, you may not have all you have, but start anyway. It is said that boldness is not the absence of fear, but to act in spite of it. Your willingness to do some good and leave a good legacy provides you with more grace. It is easier when you start to give back from your passion, helping others in what makes your heart beat. People who pursue their dream tend to give back to society more than those who were tied up to work just for money.

As you go to work today, ask yourself if, at the end of your retirement, what you know or do in the oil and gas industry will contribute in any way to the common man. Will your loved ones and entourage benefit, not just from the money and oil resources, but the same skills and time contribution? Will your children have a share, or is it all gone? How can you help? Will you join us to sponsor students and lecturers in the universities? Will you extend a helping hand to those around you? What is that thing you lacked most as you were beginning your life struggles? Could you make it available to someone starting today? Decide today to add your bit to the humanity project. The divine architect of the universe is calling.

Prepare for the Future

There is a future without you on Earth. What you leave behind and what you will become when you are no longer here is in your hands as you read this book. You might have reached the peak of science, either believing in God or being an atheist. What will it benefit you, after all the struggles, the quest for wealth and happiness, when you are gone? What will you leave behind, what will you go with? Answers to these questions give peace that you cannot get from all the struggles of life. It is the conclusive answer that brings satisfaction as you end it all.

As a mortal, one day you will leave this planet. Do you believe in God, any connection with divinity? If you agree your life on earth is temporal, you need to take care of everything you earn or have as a trust. See it as something you have the opportunity to see and take care of; something given to you by divine providence. Can you account for it, did you use it for the good of others. As you worked through the petroleum industry, was your life on earth a blessing or a curse to others? Should the almighty God call you to give an account, would you have served his purpose on earth or passed through it as if you owned everything and will never die?

These are eternal questions, which will last after your stay on earth; this world which you will pass through as someone in a foreign country. The rigs, the oil installations, the office towers and technical rooms, will not be with you forever. You are accountable for everything, including any harm to the environment and to mankind. That is why we must be pious, contrite and humble. At death, we are equalised on earth and beneath the earth. Do not live this life as if you own it; give to others, prepare for what will never end. Thinking and remembering death is part of life. Working for the afterlife is the eternal quest, for the common man as well as for the scientist; for every oil and gas worker as well. You may have it all on earth, but what happens afterwards? You will give back your soul. Ask for help if you feel a vacuum in this aspect of your life.

Afterword

Overview of the Petroleum Industry and *A Career in the Petroleum Industry* are the twin books that I hope will empower the masses and change the mindset and role of the oil and gas professionals positively. It is time to strengthen not just the petroleum sector, but the mining of other mineral resources, and use the ample resources to strengthen our ports, our tourism and agriculture. No one else will do it for us.

In a social experiment by Adrian Gee, which went viral on the internet, he pretended to be a blind person and asked for 5 dollars of change while handing over a 50 dollar note to a passerby. Most of them collected the note and cheated the fake blind man. This was not in Africa, not in Nigeria, revealing that whether white or black, humans have a corrupt tendency. It is good enforcement of laws and morals, the understanding of mutual checks that make everyone become wiser. I do not advocate hate and I insist we create an environment that is friendly for foreign investment. Let us however be wise and know that solution will come from within, back home in our own nation.

Know the business you do and play a part beyond the computer screens. Most oil and gas workers in the technical section don't even understand the economic documents they sign. I want you to know more than that; let us not just drill for oil, let us offer something to the world. Then have faith and know that only you can create that lovely future of your dreams.

Start Somewhere

When I returned to Nigeria, I did not just shout and watch. I founded a platform to contribute my own part in taking our young people from oil to true success in the energy industry. The Integrated Elvee Services Oil and Gas (iESog) stands out because of its closeness to the people, the diversification of resources, pivoting from oil and gas to agricultural biomass and then to renewable energy. It attests to the energy mix approach in solving our energy problems. We help young people to learn oil and gas technology and apply them to the development of other forms of energy. Our offices are bounded by quarter fences, not cut off from society, and we allow recreational activities not only for employees, but also for the public.

We cannot keep producing only oil, so we need to develop our own technology and do things our own way to become perfect. We should be allowed to make those initial mistakes that every other developed country or individual does before success. Technology grows and is perfected. I see how jittery our trainees become when they attempt to perform an experiment and they fail. We try to reassure them that it is only normal to attempt several times before getting it right. End users in developing nations need to know that there has to be several trials before perfection. Don't be intimidated by the final luster of imported products. Don't expect those people, who are doing their best, to get it right immediately and punish them when they don't.

For me and my team in iESog, there were challenges all the way, but the hope of the destination made every bit of the journey interesting. It was not easy to give up multimillion naira opportunities and honours tied to top executive positions in the mainstream oil company circles, to pursue a grassroots people-oriented dream in a tough economic situation with its waning oil values. The passion was stronger than the pain. That same passion should drive you today as you join this career path. I know I cannot do it alone and I am ever grateful to the team.

Using the iESog platform, we have been able to pursue a three-point agenda:

1. Engender an energy platform that is close to the people. We hold public lectures and disseminate awareness programmes about the oil industry and how the resources may be used for the betterment of society. We cannot achieve this goal if publication of energy resource material is not given a priority. We spread the message using all platforms; we have raised the standards for publishing in Nigeria and beyond through our sister company: Delizon Publishers. Delizon publishes all kinds of print and online media, but have a dedicated section on the geosciences accessible at www.delizonpublishers.com

2. Collaboration with the universities, to reposition our lecturers and students to have a direct link with oil and gas and the entire energy industry. In addition to regular trainees in our diploma programme in energy eesources, we welcome many Industrial Trainees (IT) and Students' Industrial Work Experience Scheme (SIWES) candidates annually. We train for a transition, from the petroleum as the main source of energy to renewable energy as a responsibility of the oil and gas industry. We prepare them to lead in the quest for an energy mix at national and international levels.

3. Comprehensive training and services in the industry, and expansion to the development and value-added utilisation of mineral resources and infrastructural development internally by the local people. We have become concerned not only about transition of energy, but also that of oil workers needing survival techniques for changing times, needing an exit strategy from the present to the future. By introducing them to the concept of derivative processing, we have geosciences personnel who build the equipment required to process agricultural biomass and develop renewable energy sources.

My early life as an apprentice, my experience in first degree geology, master's degrees in petroleum engineering and in geophysical exploration and reservoir, before a PhD in Sciences of the Universe, then the work experience in different domains, have opened me to a broader scope of duties cutting across disciplines. My various trainings, over fifty courses and on-the-job coaching in the industry, and my practical research, gave me the inspiration for the application of *derivative processing*. Extending this to the geosciences would mean that geologists and geophysicists are no longer limited in what they can create in a world of so many possibilities. In the iESog laboratory and workshops, and in our Biomass centre, we put things together and see what they give. Hopefully, Nigerians will begin to aim for internally developed energy solutions. Until we develop our own way of doing things, we cannot master the economics of world technological capacity building.

Go to *www.iesog.com* to find out about the training and programmes we offer. I use this opportunity to call on all concerned persons at home and abroad to give us their support and well wishes.

Send your email to: *contact@iesog.com*

iESog Trainees and IT students

Each day I see the young boys and girls who come from university to do their internship at the Integrated Elvee Services Ltd, I see hope and remembered how much I have longed to have the opportunity to invest in them – it is as if I discover their faces as the future I looked forward to. They come in many tones, fair and dark; in many ages – younger and older, and with different hopes. When we study or do the practical work, over time I see them taking on responsibilities. Those who have been with us will laugh when they see how new entries complain. We have all heard complaints and we no longer respect excuses. We just act.

One message I share is that petroleum can come from almost anything, just like energy can come from anything. We chat about petroleum, about solar and wind energy and how to simplify power distribution in Nigeria. We are imbibing the energy culture to that seed which will transcend this generation. By providing an iESog services and training centre adjoined to student's residence and corporate guest house for visiting persons, we often find ourselves in an atmosphere of oneness and unity in our pursuits. We are beginning to think knowledge, to think Africa and to think global. The iESog geosciences support for universities and other higher institutions involves technology transfer, through awareness and dissemination of technical and administrative knowledge of the energy industry within and beyond our immediate society.

Moving on to Renewable Resources

The warning about global warming is pertinent. I only ask that we remain real and make the transition to renewable energy smooth for developing countries. In a recent heart-rending video that went viral on the internet and seen by millions of people on National Geographic's website, a weak and thin polar bear was shown to be dying from the effects of climate change. Later, the film-maker revealed that it was not necessarily due to climate change, but that it could simply be an old, ill or injured bear, or starvation due to other reasons possible at any time in the world's history. National Geographic then corrected the title of the video and appologised. This is to say that many proponents of man-made climate change find a way to make people imagine a future climate disaster even when their evidence is not certain. Energy transition is everyone's concern; we do not need to make enemies or lie to achieve it.

As a contrary view, few investigators of Earth processes like Gregory Wrightstone use their geological skills to explore possible advantages of climate change. In his book INCONVENIENT FACTS, Wrightstone argues against apocalyptic predictions of climate change and says that the world may be getting better naturally. The strong point in that work is that man may not be able to control what he calls incontrollable – a natural tendency of sea level rise and fall and the trend of world climatic cycle. Persons commenting on his work emphasise that the oil industry has been horribly persecuted worldwide unjustifiably, by environmental extremists whose only goal is to bully the public into toeing their agenda in the futile attempt to control nature. Such extremists in this connotation present only climatic information that tends to support their viewpoint and look away from contrary data.

Whichever scientific side one chooses to belong, we must make the current effort to transit to renewable energy smooth, as not to destroy the good already achieved in the energy sector or ostracise a whole progress from fossil fuel. Let's support the drive to clean renewable energy; let the change we make in the industry provide capacity and help us to transit to what is better for society. Let's give this last army of personnel what they need to become pacesetters of a better world after the oil and gas era.

What our country has lacked is the long-term patience and the faith to start today. If it will take many years to create a brand of young Nigerians, we should sow the seeds today without expecting immediate rewards. This is not for politicians who harvest where they do not plant. When we train, let us know that though the techniques are important, they are not as important as the development of the able mind. That ability to try towards what you cannot see is faith and when work is added to faith, you see your achievements even before they come to existence. That is the beginning of the future. The very one that brought everything into being and without which we cannot create our own new world.

This is the new world we see in our energy platform. The Integrated Elvee Services Limited does not see petroleum only in terms of crude oil. In addition to offering services such as petroleum evaluation, traps and seal analyses and fault characterisation, we do a practical research on coal liquefaction and ethanol fuel. We cultivate plants and agricultural produce for consumption, then attempt to convert the waste to biofuel, and other forms of biomass energy, at the Elvee Biomass Centre. We have prototypes of modular refineries and crude separators, solar plates and wind turbines, built in our own workshop. We see petroleum technology as a related skill that should rightly help us to move on to better energy. We are a company in transition to renewable resources. We train and offer services in three major energy subdivisions:

1. Petroleum Geosciences
2. Agricultural Biomass
3. Wind, Solar and Power

The Integrated Elvee Services Limited headquarters is in Enugu State. Before returning to Nigeria, I considered the town of Enugu as a good resort for training in the energy sector. Enugu, known as the coal city, was Nigerian's energy town in the coal era, producing thermal and hydropower. It is a peaceful town and I can reassure my foreign friends that they can visit and not be afraid of security issues. As a low-cost town, the fees for our training are considerably reduced, making it possible for the average Nigerian. All that is needed is to provide accommodation for resident students and a guest house for visiting trainees and tutors.

Though not an oil producing State then, I was selected for my scholarship while I was in Enugu. This is why I feel every person from anywhere in Nigeria should be empowered to carry on the quest for the energy mix.

The crude resources we need for energy are available in every Nigerian State. Consider the waste dumps for example. Waste can be converted into anything and to especially solid dust, liquid power slurry and gaseous fuel. Waste can go through the process of compost like manure to be converted to liquid or gaseous fuel, while the solid can be burnt to produce syngas or direct heat for powering turbine systems based electricity.

In an interview on the Energy Platform of Nigeria Info radio, Mr Oladimeji Oresanya, former Managing Director of the Lagos Waste Management Authority (LAWMA), explained how a town can be completely powered by waste, as thousands of metric tons of waste generated per day can produce mega watts of electricity. Mr Oresanya studied geology and saw beyond just waste management but waste utilisation. The Waste Management Society of Nigeria (WAMASON) estimates that the country generates several million metric tonnes of waste per year. Ninety-nine percent (99%) of waste in Sweden is recycled, half of which is burnt to produce energy. I am strongly in support of biomass from waste as compared to that from stable food resources. This is also better than burning forest wood as a major form of biomass. When considered as waste treatment and recycling, the advantage of generating energy from waste biomass is double. Nothing should be done in excess – the reason energy mix is key. Any state of Nigeria can start generating power right away from the incineration of waste.

With the new power polices, it is now possible to generate and distribute energy as a private enterprise or as a State government. All that is needed is a licence or a Power Purchase Agreement (PPA). Independent power producers can collaborate with the Electricity Generating Companies in Nigeria (GENCOs) and the Electricity Distribution Companies (DISCOs). Users should however be able to choose their preferred supplier and switch at will to induce a positive competition of the generating and distribution companies.

I want to advocate simplicity, less strive and rather contentment in the oil and gas industry, and indeed in the entire energy sector. Time is passing and soon you and I will have passed. In the striking poem by Mario de Andrade, one of the founders of Brazilian modernism, he says, "I counted my years and realised that I have less time to live by, than I have lived so far. I feel like a child who won a pack of candies: at first, he ate them with pleasure but when he realised that there was little left, he began to taste them intensely." Towards the last stanza he says, "I am sure they will be exquisite, much more than those eaten so far. My goal is to reach the end satisfied and at peace with my loved ones and my conscience." The poem had ran for several stanzas before concluding that "we all have two lives and the second begins when you realise you only have one."

As children, we did not worry much about the future, we did our best when instructed to work and then we left the rest – allowing the future to unfold naturally. We played; we did not work from morning to night for money. As adults, after the initial struggles of life, we must start tending towards that life of a child where our world was at best our play and we did it with all our hearts and might. Again, do not compromise you inner self. As we realise our days are edging away, we become more responsible and taste our candies more intensely. That intensity is our speed at the team I lead in Integrated Elvee Services Limited.

In my other book, *Overview of the Petroleum Industry*, I mentioned how it all started when as a child I had no shoes, and wore cloths with holes, in primary school due to poverty. When events led me to have the opportunity to pursue accomplishments, I promised myself that I would help other ordinary young people in the same situation to have a chance. The pictures in that book showed how from my primary school I went to learn a trade in a photographic studio and later as an apprentice in a welding workshop in the back streets of Onitsha; all this between the ages of 11 and 15 years. Being a photographer at an early age helped me to take pictures, and catalogue most of my life history, some of which you will see at the end of this book. Grace and faith brought me back to school. I got to know about the computer, multimedia encyclopaedia and the Internet afterwards. I was already broadminded when my mentor discovered me and gave me the chance to see the world. I hope to give other people that chance today.

I moved from nowhere to attain some success, even before my scholarship and contact with the oil and gas industry. I believe other supposedly ordinary persons will pursue and meet their own dream. I will never forget the various supports I got, sometimes out of the blue, at the critical periods of my life; they overshadow the difficulties. There were and always will be some people by your side. I reminisce as far back on my days as head boy or head student in Government Science Secondary School Jalingo (GSSSJ). The head boy or head girl was tasked with many responsibilities beyond that of a regular head student in a normal school. Because of the religious riots that had occurred in the State, some of them starting from the school, I had to sit before major State panels, at that age, to give explanations on how to prevent reoccurrence and ensure stability in the school.

The head girl and head boy were the visible representatives of the students and all eyes were on us, especially when there was political tension, religious crises or tribal agitation. We had close persons, both within the students, lecturers and neighbouring institutions, who encouraged us and made us see life challenges as stepping stones to greatness. We were close to our teachers, the school and state administration, and participated in running one of the then largest secondary schools in Northeast Nigeria, in the midst of poverty, unkindly love and religious tensions. We had to resist the temptation of being incited or reacting bumptiously when external fanatics and bigots attempted to incite young students, as a way to initiate trouble in the state. It was a learning point for us as teens, to resist undue pressures and external influences. Those lessons still help me today to see and withstand the challenges in Nigeria, not letting them discourage the intention to invest and settle back at home.

I must confess that it has not been easy when it comes to collaboration with our government and exchanges with public officials. Presenting an endeavour as a free enterprise attracts all kinds of administrative taxing and interests, often a pressure on the business. We live in country where, in some states, the waste management agencies bill you in several hundred thousands of naira without ever passing by to collect your bins. We remain resolute and hope for a better playing ground. It turns out that most office holders in Nigeria, at the federal, state and local government areas, especially the appointees and some in elective offices, are not prepared to engender developmental actions in the area of homemade technological growth of our country. Most politicians have limited professional performance ratings and they hold on to the position necessary to effect a change. When you confront some of them, they explain that the situation is beyond their control, yet they do not resign.

There is no effective institution of policy making and implementation, to prepare unqualified elected or appointed officials who are compensated for their sponsorship and participation in the electoral process by their party members. We need to release the management of energy and electrical power from the government corridors to home grown private developers. The Nigerian Electricity Regulatory Commission (NERC) has decided to encourage investment in renewable energy generation among other energy types in Nigeria. They should train solution providers in societal and communal amenities as leaders and not let overwhelmed non-technocrats delay approvals as a way of showing they are important.

Thank you for reading this far, into this protracted afterword. I want to believe that this book has helped you to see what a career in the petroleum industry portends and what lies beyond it. Many pictures are included at the end to show the journey so far in this pursuit of righting the wrongs of the past in our energy industry. Henceforth, it is no longer my story, it is about you, about anyone who can now learn to generate energy in their backyard. It is the history of our nation and that of those who will make a change. As you join in this new quest, you are among those writing their own account, who will carry on the struggle for energy independence, individual freedom and that of a people. If you have not already done so, you may also obtain my other book, *Overview of the Petroleum Industry* and make an even more enthralling discovery of the oil and gas sector.

Journey so Far...in Pictures

Left: *Livinus.* **Centre** *and* **Right**: *Awo & Bros Photo Studio and the laboratory where Livinus was an apprentice in the late 80s at a tender age in Northeastern Nigeria; he was also an apprentice welder in the streets of Onitsha in the early 90s in Southeastern Nigeria.*

My early life in Northeastern Nigeria, during a time of tribal and religious conflict, exposed me to tribulations and injustice early enough. I think there was something about a vision, for fair play and social justice for the young, that early.

My childhood presented challenges and hopes despite difficulties of daily living. I left school after the primary school level to learn trades as a photographer and then as a welder. In those days, taking photographs was an art and developing them demanded even more technological skill.

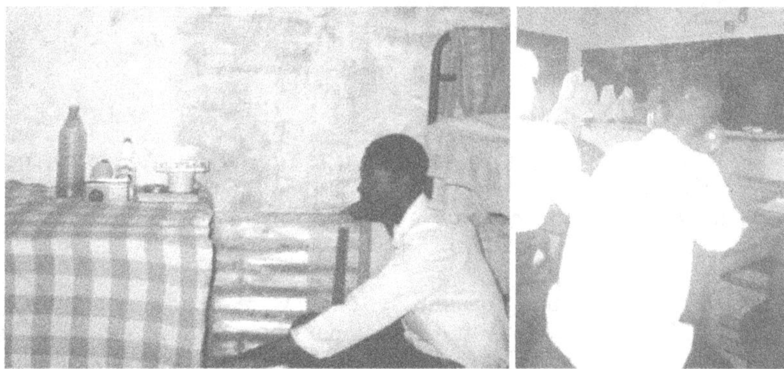

Left: *Livinus' corner during his secondary school.*
Centre: *A practical demonstration during a class session in GSSSJ.*

You needed to see my corner in the secondary school, a remnant of poverty still in reminiscence. Though located in a state classed as academically disadvantaged, Government Science Secondary School Jalingo (GSSJ) had scientific equipment and we gathered to take a look and practice during class sessions.

The Fellowship of Christian Students (FCS) executives; the set before Livinus', who eventually became the president of the association before his graduation year.

I joined FCS on admission to GSSSJ and we stood for fairness; the umbrella association provided moral support in the challenging, most notorious secondary school in Northeastern Nigeria at that time. My position as the head boy placed me in a visible role of maintaining peace among the Christian and Moslem students against external incitements by religious fanatics.

Left: *Livinus' 1995 graduating set; the arrow points to the head boy and the head girl.*
Right: *GSSSJ's agricultural activities, showing Livinus in the centre with a hoe making ridges.*

Our farm activities in secondary school were taken seriously, but very few of the students I know today continued with agriculture; many joined either the military or the police or politics. None of us was told about oil resources and petroleum technology while we were in secondary school.

I kept in touch with my University (**left**: during my matriculation); UNN lecturers influenced me, particularly two or three of them who kept in touch; we met at international professional conferences and discussed how to improve learning and tie the knot between the academia and the petroleum industry. Prof C.S. Nwajide (**Centre**: beside me) was particularly an inspiration and still is – he wrote the foreword to my technical book: *EXPLORATION AND PRODUCTION GEOSCIENCE*. Prof A.W. Mode (**Right**: handing over an award to me) remains one of those in the lead during most of iESog Annual Training and Field Trips.

Our geology field trips in Nigeria prepared me for the arduous task of climbing the Alps and the Pyrenees in Europe, in the winter and in the summer. After climbing hard, we would sit down on the peaks of the mountains and think about home. We wished to use such learning to optimise our natural resources hidden in the rocks of the basins and valleys in Nigeria.

I compare our geological fieldwork today with those in my university, first in the Anambra Basin and later in the Alps and Pyrenees, and the mix of cultural programmes for youth exchanges while learning in Europe. I synch the differences among them and make provisions so iESog trainees will benefit from all there is to experience.

One way to not lose touch, while I was away, was keeping close to my alma mater; I visited the university from Europe for lectures and rekindled my relationship with the NAPE and AAPG chapter of the school, where I got an award as the founding president of the association.

I particularly thank God for the aspect of the Mass Internet Literacy Campaign (MILC), which we carried out about 15 years ago in Bayelsa State. Some of the participants were children, but when they contact me on the social media today I find adult young men and women achieving great things in society.

We started with one computer, but eventually had several and enough to donate to the NYSC for public enlightenment. As I tried to donate to others, things got better for me. Life did not continue to be difficult – all through, a new life tended to make me forget the past.

As I see myself among these children in the picture, I remember that I too was young then and had faith that we would do more. God answers prayers – anyone who believes and works in faith will have his desires met.

God answered my prayers and gave me all I had wanted when I got the opportunity to be trained in Europe. I will not fail to do the very mission I am called to; I want this same pact extended to others who will keep up the efforts to individually contribute and make a difference for our society.

Park Valrose, and the *Château de Valrose*, my school of doctoral studies castle area, at the University of Nice, has a wonderful garden that can make you forget your past!

I played hide and seek in that park with my lovely wife (standing with me in the picture in front of University of Nice, Valrose campus) and we almost, but thankfully, never forgot our pasts.

Nice town, on the French Riviera – the Mediterranean, in southeastern France, is a very lovely town. The whole of the *Côte d'Azur*, including the popular city of Cannes, was our stopover to visit the restaurants and the beaches, to see and feel life. With these good days in Europe, and starting to work in France, I had the opportunity to never return again to Nigeria. Some of the persons I left Nigeria with are never coming back home to face the challenges of our nation at its grassroots.

A cross-section of team-building events during Livinus's training and work in Europe

Career profiling even out of office could change one; for me, life in Europe was different – an icing of sugar amidst undue internally non-voiced pressure to succumb.

I have tried everything, in the mountains and in the valleys, in the snow and on the beaches, in space centres and in theme parks, but in all this, I choose to never forget where I am coming from – my history which carries me along the journey so far.

The promise I made to myself brought me back to Africa when my life was being carried away by the winds of pleasures and pressures in Europe. I had to give them up to continue the grassroots development project among Nigerian young people from where I left it.

We continue the grassroots training today, from the basic computer and internet literacy campaign of those days, to the geological models and advanced petroleum geosciences with all-round energy activities.

Group photos during field trips in the Anambra Basin, to rediscover its petroleum and unconventional resource potentials, as well as energy-related economic minerals.

We make it available for beginners as well, giving them the chance to make a living from agriculture and food processing, while harnessing the abundant energy resources from biomass.

Left: *Integrated Elvee Services Limited Main Office*
Centre: *Elvee Agricultural Biomass Centre*
Right: *Elvee Renewable Energy Section*

Our activities embody energy mix: petroleum, biomass, wind and solar. This includes the basic energy need by man in the form of food. The iESog Agricultural Biomass centre serves for food processing, and the wastes are collected for recycling or conversion to energy.

In iESog residences and training centres, the trainees are encouraged to be in high spirits; we believe that is the height from where innovation comes. The work and play method applies.

Left: *Trainees and IT students in the Elvee Biomass Centre*
Centre: *IT students in the iESog Workshop*
Right: *Hydrocarbon Fluorescence Test in the iESog Laboratory*

The trainees are advised to spend time in the iESog laboratory and workshop, and practice with equipment and sample resources, to create harmony between them and the realms of creativity for new results.

Our training is only complete when the students are adequately exposed to class work, laboratory practicals and operational experience in the field.

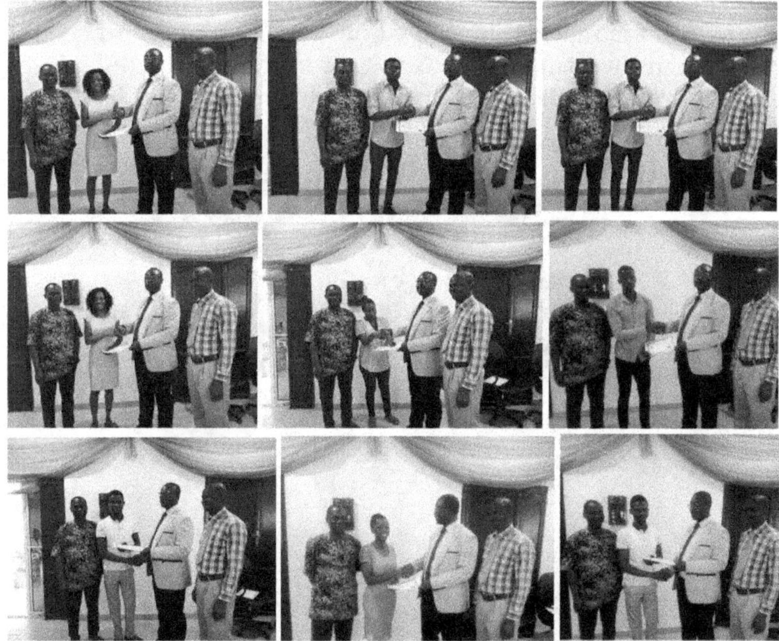

Certificate award sessions with Prof A.W. Mode and Dr Charles Ugbor

Congratulations to the different sets of iESog trainees; every candidate from the diploma programme in petroleum geosciences and energy resources cannot be represented here – the few pictures of certification suffice as a pacesetter for future candidates who will share from the opportunity.

Also by the Author

Overview of the Petroleum Industry reflects a picture of the author's pact with the people of Nigeria, in which he highlights the good, the bad and the ugly sides of the oil and gas industry, revealing why the system that controls resources and cash flow, a supposed heritage, will do and has been doing everything to sustain itself. The work is an exposé into the nitty gritty of the oil and gas world in the ordinary man's language, enabling the reader to know the organisation of the sector, from the exploration and production processes to finance.

The book probes the petroleum sector from its core profile and covers issues on health, safety and the environment. The odious process of contracting and procurement, and petroleum marketing, show how misdirection of proceeds from foreign investments leads to non-sustainable economic trends in developing countries of oil-producing regions.

About the Author

Livinus Nosike leads the first indigenous fully-integrated petroleum geoscience training and services, the Integrated Elvee Services (IES), with a mindset of equipping everyone with a comprehensive knowledge of the oil and gas industry and the energy sector in general.

The Integrated Elvee Services Limited is currently made up of the Petroleum Geoscience Services unit, the Agricultural Biomass centre and the Renewable Energy section section. IES Oil and Gas (iESog) offers a synthesis and synergetic energy information base, an exploration-level petroleum assessment and oil well operations support. iESog extends its hand of fellowship to universities and other tertiary institutions all over the country through its *Energy Supports Consortium.*

The author obtained the Talent Development Scholarship Award in 2004 for his post-graduate studies in Europe sponsored by the TotalFinaElf (Operator of the Joint Venture with NNPC) and the French Government. On completion of the programme, he worked at Total's Technical headquarters, Jean Feger's Technical and Scientific Centre (CSTJF), in France for a few years before returning to work for Total E&P Nigeria (TEPNG). He underwent operational and on-the-job training in Aberdeen, the North Sea, Norway, East Africa, and Nigeria. He has worked as a wellsite, operations, exploration and synthesis geologist and as a geoscientist.

Livinus holds a BSc in Geology from the University of Nigeria Nsukka (UNN) and Master's degrees in Petroleum Engineering and Geophysical Exploration & Reservoir, from the University of Pau (UPPA), France. He obtained a PhD in Sciences de la Terre et de l'Univers

(Sciences of the Earth and of the Universe, equivalent of Earth and Planetary Sciences) with the highest and most prestigious doctoral honours in France: *Très honorable avec les félicitations orales du jury*, from the University of Nice – Sophia Antipolis. One of his case studies was the relationship between tectonics and hydrocarbon leakages in the deep offshore Niger Delta.

He founded iESog after consultations with industry players and regulating authorities in the sector. As an expert in the upstream sector, he drives strategic development through empowerment programmes from which many individuals and corporate organisations have benefited. He has written and presented several papers at local and international conferences as well as authored five petroleum geoscience technical books, currently in the publication chain of the Petroleum Geoscience Book Series by Delizon Publishers.

His book, *Overview of the Petroleum Industry*, targets the general masses concerned about the challenges of the petroleum industry, especially in Nigeria. His latest technical book is titled, *EXPLORATION AND PRODUCTION GEOSCIENCE – for a Comprehensive Skills Acquisition in a Time of Energy Transition*. The book puts together what geosciences candidates need to know, covering the full petroleum chain from exploration through appraisal to development and production, in one volume, aiming to provide a synthesis that guides the reader to be skilled in the whole. It is a must read for would-be and existing professionals during a fossil-fuel *mop-up exercise* in an evolving industry.

The author is a member of the Nigerian Association of Petroleum Explorationists (NAPE), the American Association of Petroleum Geologists (AAPG) and the Nigerian Mining and Geosciences Society (NMGS). He was the founding president of both NAPE and AAPG chapters in 2000 at the University of Nigeria Nsukka

(UNN). He is also a cofounder and trustee of the UNN Geology Alumni.

He has sponsored many educational programmes, including the iESog and the Delizon Scholarship Schemes and Writing Competition. He is the convener of the iESog Annual Training and Field Trip in the Lower Benue Trough, involving sponsorship for university lecturers and students. He is interested in education for the masses and was the pioneer of the award-winning Mass Internet Literacy Campaign (MILC) during his National Youth Service Corp (NYSC) programme in Bayelsa State. He has authored books on academic success and on Internet literacy for the 21st century.

9 782365 230858